Harold Whiting

A Course of Experiments in Physical Measurement

Vol. 1

Harold Whiting

A Course of Experiments in Physical Measurement
Vol. 1

ISBN/EAN: 9783337337377

Printed in Europe, USA, Canada, Australia, Japan

Cover: Foto ©berggeist007 / pixelio.de

More available books at **www.hansebooks.com**

A

COURSE OF EXPERIMENTS

IN

PHYSICAL MEASUREMENT.

In Four Parts.

PART I.

DENSITY, HEAT, LIGHT, AND SOUND.

By HAROLD WHITING, Ph.D.,

FORMERLY INSTRUCTOR IN PHYSICS AT HARVARD UNIVERSITY.

SECOND EDITION.

BOSTON, U.S.A.:
D. C. HEATH AND CO., PUBLISHERS.
1892.

PREFACE.

THIS book is intended to aid in the preparation of students for courses in civil, electrical, or mechanical engineering, and for advanced work in all branches of science requiring the use of accurate methods and instruments of precision. To this end, the course of experiments described, unlike that contained in most manuals published in this country, is exclusively devoted to quantitative physical determinations. Comparatively little use is made of the ordinary experimental demonstrations of well-known physical laws and principles, which, it is believed, are better suited to the lecture-room than to the laboratory. Most of the experiments consist in the determination of magnitudes wholly unknown to the students, and are made with instruments which they themselves have tested, in order that they may learn to depend upon their own observations.

Attention has been paid throughout this book to the *general methods* which underlie all physical measurement, rather than to the special devices by which particular difficulties are overcome. It is considered of greater advantage to show how comparatively accurate measurements may be made with rough apparatus, than to explain the use of instruments of precision which, in the hands of students, are apt to give erroneous results. The apparatus required for this course is, accordingly, of the simplest possible description.

Most institutions are obliged, by considerations of expense, to limit either the quantity or the quality of the instruments provided for the laboratory. When the supply of apparatus is insufficient, the work of a given student at a given point of time is obviously determined, to a greater or less extent, by the instruments which happen to be free for him to employ, and the systematic instruction of large classes becomes impracticable. This book is intended especially for use in laboratories which are or can be provided with a liberal supply of moderately accurate apparatus. Effort has been made to devise inexpensive instruments, especially when several copies of a given kind are likely to be needed; and it has been found that, notwithstanding the expense of all necessary reduplications, a considerable saving may be effected by the "Collective System" of instruction in the cost and labor of conducting an elementary laboratory course. The experiments are accordingly such as can be once for all explained to, and within a reasonable length of time performed by a large class of students. They are moreover arranged in a connected and progressive order.

The care and accuracy required to obtain concordant results in Physical Measurement, the continual use of experimental, inductive, and controlled methods, give to that science a peculiar educational value, aside from the natural laws and principles with which the student must become familiar. The course of experiments has been adapted, in so far as possible, to the needs of students who, having little or no previous training either in mathematics or in physics, wish to obtain a general scientific education. Every branch of physics is accordingly represented by typical examples. In order, however, not to exceed the natural bounds of an elementary treatise, the author has limited his selection to such experiments as have been proved, practically, in his own experience, to yield the most satisfactory results from an educational point of view. It is hardly necessary to add that these experiments involve physical measurement in every case.

The amount of mathematics required in the use of this book is not so great as might be supposed from a casual examination of its pages, since many proofs are given in full which in other text-books are taken for granted. The course of one hundred experiments involves only the simplest propositions in arithmetic and geometry, and little or nothing of algebra or trigonometry beyond the mere notation. Problems presenting any special difficulty are treated separately in a portion of the Appendix (Part IV.) not intended for general use.

The first part of this book relates especially to hydrostatics, thermics, optics, and acoustics; containing measurements of mass, density, length, temperature, heat, light, and wave-lengths of sound.

The second part contains all such measurements as involve motion or acceleration. That part of acoustics which relates to the measurement of time is also included; then follow dynamics, magnetism, and a comparatively extended series of electrical measurements. A few experiments intended (with certain exceptions) for advanced students are added, together with a description of certain instruments of precision.

The third part contains notes on the general methods of physical measurement, and on physical laws and principles. An extended series of mathematical and physical tables is also included in this part.

The fourth part, or Appendix, contains suggestions to teachers in regard to laboratory equipment, apparatus, expenses, and methods of instruction. It includes a full set of examples, showing how the observations in the course of one hundred experiments should be recorded and reduced. These examples embody results a great part of which were actually reported by students. There are also three working lists of experiments, of different lengths and degrees of difficulty, and proofs of certain important mathematical formulæ.

The text of the first and second parts is divided into short chapters, distinguished by the names of the experiments (Exps. 1–100) to which they relate. The experiments are still farther divided into sections (¶¶ 1–270), devoted in some cases to the practical, in other cases to the theoretical treatment of the subject. It has not been thought necessary or desirable to indicate in all cases just what portions of an experiment the student is expected to perform, and what portions it is sufficient for him to read. This must, of course, depend largely upon circumstances. Full directions for each of the one hundred regular experiments, or for each part of which it consists, will usually be found in a separate section headed by the word "Determination." In the case, however, of outside experiments mentioned only for the sake of illustration or continuity, directions are either entirely omitted, or replaced by a mere outline of the methods involved, with which it is important that the student should become acquainted. Examples will be found under the "Peculiar Devices employed in Calorimetry" (¶ 97), and the "Velocity of Light" (¶ 247), which, though obviously impracticable, even for advanced students, furnish reading matter which is none the less instructive.

More than half of the sections in the first and second parts relate to principles involved in the experiments, the construction of the necessary apparatus, or the calculation of results. These should be read or omitted by the student at the discretion of the teacher. The references to the third part (§§ 1–156), which occur throughout the experiments, should be looked up by the student in the order in which they are met, and afterward read consecutively. The teacher should make sure that these references are understood, in the case especially of students who may have had no previous training in physics.

The examples in the fourth part are intended to aid the teacher in preparing a list of the data required for a given determination, and in explaining the reduction of these data. The calculations are made, for the most part, by purely arithmetical processes, and in so far as possible, by one step at a time, so that the student can hardly fail to understand them. The author has found in his own experience that such examples can be safely trusted in the hands of students; but, for obvious reasons, it was thought better that they should be contained in the fourth part or Appendix, copies of which, separately bound, can be used by teachers who prefer to keep the examples at certain, or at all times, in their own hands.

The three lists of experiments, proposed by the author with a view of preparing students for various requirements of Harvard College, may be useful also to teachers who wish merely to shorten the course of experiments described in this book, without interrupting the continuity of the course.

The mathematical portions of the Appendix contain proofs which may be of interest to ambitious students and a convenience to teachers who find it desirable to step *beyond the limits* of this book.

Few references are given to works of other authors. It has been thought better in an elementary book to incorporate in the text such abstracts from the best authorities as it may be necessary for the student to refer to. The course of experiments here described was elaborated from one previously given by Professor Trowbridge, and outlined in his "New Physics" (Appleton, 1884). In this course frequent reference was made to the well known works of Everett, Kohlrausch, and Pickering. It is impossible to say to what extent the author may be indebted to these sources for the ideas contained in this book.

The advanced sheets of a " Syllabus " of experiments arranged by the author were distributed to his class in the year 1884–1885, before the works of Glazebrook and Shaw, and Stewart and Gee, could be obtained. While considerable assistance was derived from these works in the preparation of this book, the " Syllabus " mentioned above was taken as the basis for most of the experiments. The notes contained in the third part were first distributed to students in 1888–1889, but largely rewritten in 1890. The tables were condensed, by permission, from those of Professors Landolt and Börnstein, and from other sources elsewhere acknowledged. The first part was printed in 1890; the remaining three parts in 1891. In the same year a corrected edition of the first three parts was prepared for the use of students, and all four parts were combined in a single volume for the use of teachers and students.

The author is indebted to Professor Trowbridge for an outline of many successful experiments; to Professor Hall for a revision of a part of the proof-sheets, for numerous useful and practical suggestions, and for parts of experiments taken from his elementary course; to the late Mr. Forbes, of the Roxbury Latin School, for important criticisms; and to Mr. Edgar Buckingham, Assistant in the Jefferson Physical Laboratory of Harvard University, for valuable aid in preparing the course of experiments.

The author wishes also to acknowledge several errata kindly pointed out to him in earlier copies, and to state that he will gladly receive from any source further corrections or criticisms which may be of service in preparing a revised edition of this book.

CAMBRIDGE, November, 1891.

TABLE OF CONTENTS.

Part First.

MEASUREMENTS RELATING TO

DENSITY, HEAT, LIGHT, AND SOUND.

PRELIMINARY EXPERIMENTS.

PAGE

I. DIRECT MEASUREMENT OF DENSITY 1

Nicholson's Hydrometer.

II. TESTING A HYDROMETER 8
III. WEIGHING WITH A HYDROMETER 13
IV. WEIGHING IN WATER WITH A HYDROMETER . . 14
V. ATMOSPHERIC DENSITY 17

The Balance.

VI. TESTING A BALANCE 27
VII. CORRECTION OF WEIGHTS 38
VIII. WEIGHING WITH A BALANCE 42

DENSITY.

DENSITY OF SOLIDS.

		PAGE
IX.	THE HYDROSTATIC BALANCE, I.	43
X.	THE HYDROSTATIC BALANCE, II.	46

The Specific Gravity Bottle.

XI.	CAPACITY OF VESSELS	49
XII.	DISPLACEMENT, I.	53
XIII.	DISPLACEMENT, II.	56

DENSITY OF LIQUIDS.

XIV.	DENSITY OF LIQUIDS	58
XV.	THE DENSIMETER	59
XVI.	BALANCING COLUMNS	63

DENSITY OF GASES.

| XVII. | DENSITY OF AIR | 67 |
| XVIII. | DENSITY OF GASES | 69 |

LENGTH.

XIX.	MEASUREMENT OF LENGTH	71
XX.	TESTING A SPHEROMETER	83
XXI.	CURVATURE OF SURFACES	88

HEAT.

COEFFICIENT OF EXPANSION.

		PAGE
XXII.	Expansion of Solids	90
XXIII.	Expansion of Liquids, I.	94
XXIV.	Expansion of Liquids, II.	101

TEMPERATURE.

XXV.	The Mercurial Thermometer	104
XXVI.	The Air Thermometer, I.	119
XXVII.	The Air Thermometer, II.	127

CHANGE OF PHYSICAL CONDITION.

XXVIII.	Pressure of Vapors, I.	132
XXIX.	Pressure of Vapors, II.	135
XXX.	Boiling and Melting Points	140

CALORIMETRY.

XXXI.	Method of Cooling	144
XXXII.	Thermal Capacity	157
XXXIII.	Specific Heat of Solids	178
XXXIV.	Specific Heat of Liquids	184

Latent Heat.

XXXV.	Latent Heat of Solution	194
XXXVI.	Latent Heat of Liquefaction	199
XXXVII.	Latent Heat of Vaporization	202
XXXVIII.	Heat of Combination	205

RADIATION.

		PAGE
XXXIX.	RADIATION OF HEAT	212

LIGHT.

XL.	PHOTOMETRY	222

FOCAL LENGTHS.

XLI.	PRINCIPAL FOCI	230
XLII.	CONJUGATE FOCI	236
XLIII.	VIRTUAL FOCI	239

OPTICAL ANGLES.

XLIV.	THE SEXTANT	244
XLV.	PRISM ANGLES	255
XLVI.	ANGLES OF REFRACTION	257
XLVII.	WAVE-LENGTHS	264

SOUND.

XLVIII.	INTERFERENCE OF SOUND	270
XLIX.	RESONANCE	272
L.	MUSICAL INTERVALS	273

PHYSICAL MEASUREMENT.

Part First.

MEASUREMENTS RELATING TO DENSITY, HEAT, LIGHT, AND SOUND.

EXPERIMENT I.

MEASUREMENT OF DENSITY.

¶ 1. **The Density of a Rectangular Block.** — The volume of a rectangular block may be defined as the product of its length, its breadth, and its thickness. If, accordingly, each of its three dimensions has been measured (§ 1) in centimetres (§ 5), we may find the volume of the block in cubic centimetres by multiplying these three dimensions together. When two blocks are of exactly the same size, but of unequal weight, as for instance a block of wood and a block of metal, they are said to differ in respect to density. Obviously, to determine the density of a body, we must find its weight as well as its volume. For convenience in calculation, the weighing should be made in grams (§ 6), since density is customarily expressed in grams per cubic centimetre (§ 9). To calculate the density of a body, we divide its weight in grams by the number of cubic centimetres contained in its volume, and thus find the weight of one cubic centimetre. This is the density (or average density) in question, expressed in absolute units of the C.G.S. system (§ 8). It should be noted that in this system *the density of a body is equal to the weight in grams of a cubic centimetre of the substance of which it is composed.*

The density of a fluid cannot, for obvious reasons, be determined like that of a solid, by *direct* measurements of its weight and linear dimensions; but when the volume of a block has been found, there are various methods by which the weight of an *equal bulk* of a fluid may be determined. We may, for instance, find the weight of the fluid necessary to fill a mould or vessel into which the block exactly fits; or we may fill a vessel with the fluid, and weigh the quantity which runs over when the block is immersed; or we may load the block [1] until it neither floats nor sinks in the fluid, — the weight of the block being in this case equal to that of an equal bulk of the fluid (§ 64). Other methods will be described in experiments which follow. The density of a fluid is always calculated, like that of a solid, by *dividing its weight by its volume.* We have seen how one may find the *weight* of a certain quantity of a fluid equivalent in volume to a rectangular block; the *volume* of the fluid in question (being equal to that of the block) is calculated by multiplying together the length, breadth, and

[1] In a wooden block, auger-holes bored parallel to the grain may be nearly filled with lead, and closed with a wooden plug even with the surface. A cube measuring 10 cm. each way and weighing 998 g. will be found useful to illustrate the density of water. The block should be coated with oil or other material impervious to water.

thickness of the block. All measurements of density will be found to depend more or less directly upon linear dimensions as well as upon weight.

The density of water may be found, approximately, by any of the methods suggested above; but the exact measurement of the density of water is one of the most difficult problems in physical measurement. We shall need continually to refer to the values in Table 25, which have been obtained by combining the results of the most careful observers. The student will of course accept these values in preference to any which he himself may obtain; but to use them intelligently, he must thoroughly understand both what they represent and how they are found. He should convince himself that the density of water is not far from unity; or that, in other words, 1 *cu. cm. of water weighs nearly* 1 *g.* (see § 6); and he should familiarize himself with the fundamental method of measuring density by weight and linear dimensions, applicable, as we have seen, either to a solid or to a liquid.[1] In case that a rectangular block is used, the necessary data are its weight in grams, and its length, breadth, and thickness in centimetres. The observations are made as stated below.

¶ 2. **Determination of Weight**   Fig. 1.
by the Method of Trial. — The block is to be weighed with rough scales, such as are represented in Fig. 1, and which should be affected by a decigram. To

---
[1] See the Harvard University List of Chemical Experiments, Exp. 1.

select the weights necessary to balance a given body requires in general many trials. The number of trials may be greatly reduced, in the long run, by a strict adherence to the method here described. (See § 35, 2d ed.) We first place the block on one scale-pan, and a single weight, which we judge to be nearly equal to it, on the other. If this weight is too small, that is, if it is insufficient to lift the block, we add to it another weight of about equal magnitude, if any such exist in the set of weights; or should there be no weight equal to the first, we add one of the next greater magnitude. If the two weights together fail to lift the block, we add a third as nearly equal to the sum of the other two as may be convenient, and thus by doubling the weight in one scale-pan as many times as may be necessary, we find a quantity capable of lifting the load in the other scale-pan. If on the other hand, the first weight tried lifts the block, that is, if it is too heavy, we substitute for it one half as great, if any such be contained in the set; otherwise, the largest weight less than half of the first; and if the second weight is too great we substitute in the same way a third weight not greater than half of the second, and so continue to halve the weight until finally it is lifted by the block.

The weight of the block thus becomes known between two limits. We next try a weight as nearly half-way between these limits as may be obtained by the addition or subtraction of one weight at one time, or by the substitution of one weight for another; and thus gradually approximate to the weight of the block by successively halving the interval between the limits known to contain it.

By aimless departures from this method of approximation, the number of trials may be indefinitely increased; but certain modifications may be advisable when, from the slow motion of the scales or from any other cause, one has good ground to think that the true weight has been nearly found. In all such cases one should add or take away only so much weight as may be reasonably expected to turn the scales.

When the block has been exactly counterpoised by weights, it should be transferred to the other scale-pan and balanced against the same weights as before. (See § 44.) If the scales are as accurate as they are "precise," (§ 48, 2d ed.) the equilibrium will not be disturbed, otherwise a readjustment of the weights will be necessary. In the latter case the average of the two weighings is adopted as the true weight of the block. (See Experiment 8.)

¶ 3. **Determination of Length, Breadth, and Thickness by a Vernier Gauge.** — We have seen in ¶ 2 how the weight of a block can be found; it remains to measure its length, breadth, and thickness, in order that its density may be determined. A Vernier gauge (Fig. 2) is suitable for this purpose. To obtain great accuracy with such a gauge, special precautions are necessary (see Experiment

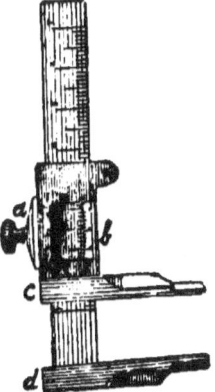

Fig. 2.

19). For the purposes of this experiment, however, it will be sufficient to observe that the distance between the jaws (c and d) is directly indicated on the main scale of the instrument by the "pointer" or "zero" of the Vernier scale (b) on the sliding piece ($a\ b\ c$), to which the jaw (c) is attached. To identify the zero of the vernier, we bring the jaws (c and d) into contact; the zero of the vernier should then come opposite to the zero of the main scale. For convenience in reading the vernier, the zero is

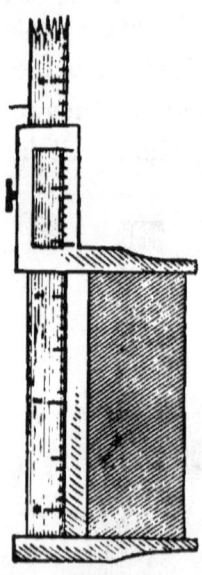

Fig. 3.

generally placed at a point (b) considerably beyond the movable jaw (c); but if, as in the figure, the main scale begins at an equal distance from the fixed jaw (d), the readings will not be affected. Evidently, in such a gauge, the edge of the sliding jaw (c) cannot be used as an index.

The whole number of millimetres between the jaws is equal to the number of the first millimetre division below the zero of the vernier, that is, between it and the zero of the main scale. The tenths of millimetres above this whole number may be read from the vernier as explained in § 40.

The block is first clamped lengthwise between the jaws of the gauge as in figure 3, and ten measurements are thus taken at different points. It is then clamped so as to obtain in a similar manner ten measurements of its breadth, and finally ten of its thickness. In each case the readings are made to millimetres and tenths. The object of taking a large number of measurements is to find the *average* length, breadth, and thickness with a degree of exactness (§ 48) corresponding to that attained in the weighing already performed (¶ 2). We finally calculate the volume and density of the block as explained in ¶ 1.

¶ 4. **Corrections Disregarded in Experiment 1.** — The vernier gauges which we usually employ are supposed to read correctly at 0° Centigrade; and hence will not be quite accurate at ordinary temperatures. For instance, if the gauge, having been cooled by melting ice to 0°, is fitted to the block as in Fig. 3, then allowed to become warm through contact with the air of the room, it will no longer fit the block as closely as it did, owing to expansion of the metal by heat. The block, though really unchanged in size, will appear to be somewhat smaller than before. This effect of expansion is barely perceptible; but we tend, nevertheless, to underestimate all the dimensions of the block, and hence also its volume. With brass gauges at 20°, the error in the volume would amount to about 1 part in 900 (see Table 8 *b*, also § 83).

Another source of error lies in the fact that the weighings are made in air, and not *in vacuo* (§ 65). In the case of a body weighing about one gram to the cubic centimetre, it is found (see Table 21), that the atmosphere exerts a buoyant action which apparently deprives it of about one 900th of its weight. We should therefore underestimate both the weight and the volume, in such a case, in the same proportion; and the density obtained by dividing the one by the other would not be affected. Even when the corrections in this experiment do not, as above, completely offset one another, they generally amount to less than one part in a thousand, and may be neglected in comparison with errors of observation. (See § 24.)

EXPERIMENT II.

TESTING A HYDROMETER.

¶ 5. **Determination of the Sensitiveness of a Hydrometer.**—A Nicholson's hydrometer is to be loaded as in Fig. 4, by placing weights in the upper pan, a, until a small ring round the lower part of the wire stem sinks just beneath the surface of the water; then small weights are added, say 5 centigrams, until by the sinking of the instrument, another ring round the upper part of the stem is brought just below the water level. The dis-

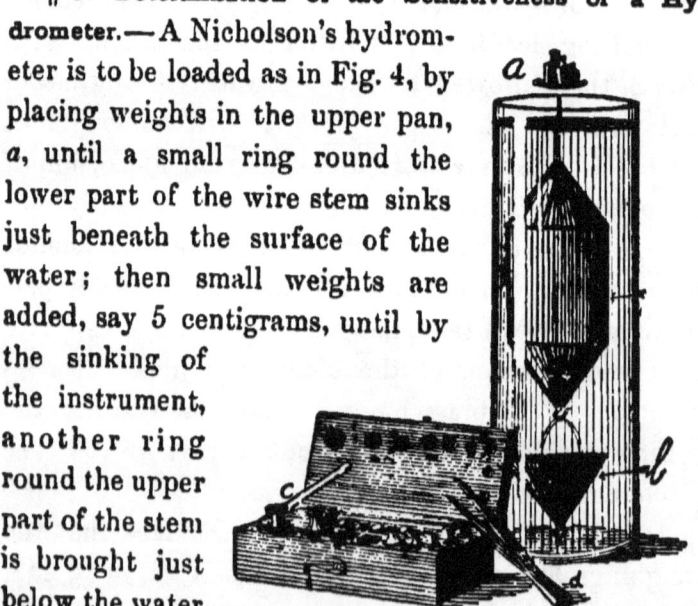

Fig. 4.

tance between the two rings, through which the hydrometer sinks under the action of the weight added, is estimated roughly by a small millimetre scale. We now calculate the effect of one centigram in sinking the instrument. This is called the sensitiveness (§ 41) of the hydrometer, and is useful in determining the degree of precision with which the adjustments of the instrument should be made (see § 48). Thus if the effect of one centigram is distinctly perceptible, we should try to avoid errors even less than a centigram in magnitude.

In using a Nicholson's hydrometer, several precautions should be observed. It frequently happens that through friction against the sides of the vessel, or through capillary phenomena where the surface of the water meets the stem, the hydrometer is unaffected by any slight change in the load. To avoid the first difficulty, the instrument should be kept floating in the middle of the jar, by the use of a guide of some sort. Such a guide may be conveniently constructed of wire, as in Fig. 5. To avoid the uncertainty of capillary action, the stem of the hydrometer should be kept wet, by a camel's-hair brush, for at least a centimetre above the water level.

FIG. 5.

In water freshly drawn bubbles of air are apt to form, clinging to the sides of the hydrometer. These should be removed by the same brush. The formation of air bubbles may generally be prevented by using either distilled water, or water which has been standing for some time in the room.

It is important to keep the upper part of the stem, the pan, and the weights absolutely dry. The guide (Fig. 4) should prevent the hydrometer from sinking completely below the surface.[1]

¶ 6. **Accurate Adjustment of a Nicholson's Hydrometer.** — A mark is made near the middle of the stem of the hydrometer and the load is altered, a centigram at a time, until this mark is floated as nearly as possible in line with the surface of the water. If

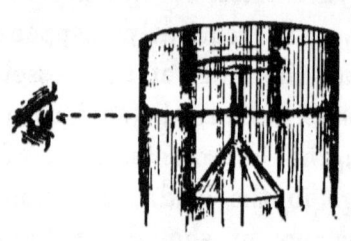

Fig. 6.

a glass jar is used, it is better to sight this mark by the under surface of the water, as shown in Fig. 6.

In the absence of weights smaller than one centigram, we estimate and record fractions of a centigram as follows: when the mark is floated exactly on a level with the surface of the water, the fact is expressed by placing a cipher in the third decimal place (belonging to the milligrams). If, however, a given weight fails to sink the mark to this level, while the addition of one centigram sends it as much below the surface as it was before above it, half a

---

[1] A sheet of cardboard or metal with a hole in the middle, is recommended by some authorities (see Pickering's Physical Manipulation, Article 45) to serve as a guide, and at the same time to prevent the weights from falling into the water. A student relying upon this safeguard is apt, however, not to acquire a sufficient degree of skill to prepare him for the manipulations of a delicate balance. (Exps. 6-14.)

centigram or 5 milligrams is obviously the weight to be added; hence the original weight should be followed by a 5 instead of a 0 in the last place. Thus if with 25.99 g. the mark is 2 mm. above the surface of the water, and with 26.00 g. it is 2 mm. below it, the weight sought must be 25.995 g. Again, if the lesser of two weights differing by one centigram is evidently nearer than the other to the weight desired, we substitute a figure 2 or a 3 for the 5 in the last place, or if the greater weight is more accurate, we write a 7 or an 8 instead. Any distinct information of this kind should always be recorded when possible, by means of a figure in the last place, even if that figure be extremely doubtful (§ 55). Closer estimates will hardly be justified in the case of a Nicholson's hydrometer.

¶ 7. **Effect of Temperature on a Nicholson's Hydrometer.** — The temperature of the water in the jar is now taken. The water is then cooled with ice to about 10°, and the weight required to balance the hydrometer is determined as before, with a new observation of temperature. Then the jar is filled with tepid water (at about 30°) and the experiment is repeated. A comparison of the different results shows how much the buoyancy of water is affected by temperature. For this purpose the observations which we have now obtained at three different temperatures are to be represented graphically on co-ordinate paper by three points, A B and C, as explained in § 59, and through these points the curve $A\ B\ C$ is to be drawn with a bent ruler. (See Fig. 7.)

The ambitious student may supplement this experiment by using water hotter than 30° and colder than 10°, also water at intermediate temperatures. He will thus obtain data for plotting a more com-

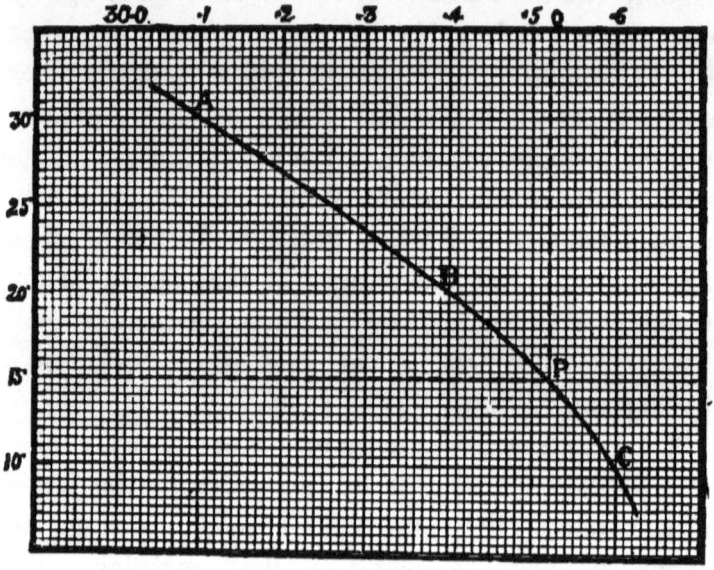

FIG. 7.

plete curve than that shown in the figure. This curve, if we neglect the expansion of the metal of which the hydrometer is composed, represents the relative buoyancy, or (see § 64) the relative density, of water at different temperatures.

EXPERIMENT III.

WEIGHING WITH A HYDROMETER.

¶ 8. Determination of Weight in Air by a Nicholson's Hydrometer. — From the results of Experiment 2 it is possible to find (see § 59) the weight necessary to sink a hydrometer to a given mark in water of any ordinary temperature. It is obvious that in all determinations with a Nicholson's hydrometer, the temperature of the water must be observed at the time of weighing. To find the weight of a body, place it in the upper pan (a, Fig. 8), and with it enough weights from the box to sink it to the same mark as before. Evidently less weight will be required than at the same temperature without the body, and the difference will be equal to the weight of the body in question.

Fig. 8.

¶ 9. Reasons for Neglecting Corrections for the Buoyancy of Air. — Since the air buoys up both the brass weights and the body used, the result of this experiment is what we call the apparent weight of the body in air (§ 65). The amount of this buoyancy depends (see § 68) in one case upon the density of the brass weights, in the other case, upon that of the body in question; hence if these two densities are approximately equal, the air will exert nearly the same force in both cases. The result, obtained as we have seen by difference, will not therefore be affected to an appreciable extent.

For the purposes of this and other experiments which follow, we choose ten steel balls, perfectly round and uniform in size, such as are used in the bearings of the front wheel of a bicycle. The density of these balls (7·8) is not far from that of the brass weights (8·4), and it will be seen by reference to Table 21 that the correction for the buoyancy of air may be wholly disregarded.

EXPERIMENT IV.

WEIGHING IN WATER WITH A HYDROMETER.

¶ 10. **Determination of Specific Gravity by a Nicholson's Hydrometer.** — The steel balls used in the last experiment are now to be placed in the lower pan of the hydrometer (l. Fig. 9), which is lifted by the stem out of the water for this purpose. The instrument is then balanced with weights, and the temperature of the water observed as in the last two experiments.

In lowering the hydrometer into the jar, care must be taken to remove with a camel's-hair brush all bubbles of air from the steel balls, as well as from the sides of the hydrometer, and also, of course, not to spill any of the balls. In the adjustment of weights the same precautions must be used as in the last two experiments. We have already obtained the weight of the steel balls in air (¶ 8); we find similarly their weight in water from the results of Experiments 2 and 4, and finally their apparent specific gravity (see § 66).

¶ 11. **Use of the Methods of Substitution and Multiplication.** — It will be noted that in Experiment 3 the unknown weight of a body takes the place of a known weight of brass used in Experiment 2; the one is in fact substituted for the other. The method of finding the weight of a body by a Nicholson's hydrometer is therefore essentially a method of substitution (§ 43). This statement also applies to the determination of weight in water by the same instrument; for the weight of a body in water is here substituted for a known weight of brass in air. The errors committed with a Nicholson's hydrometer depend upon the peculiarities of the instrument itself, rather than upon the quantities weighed. We are in fact liable to the same error in weighing one bicycle ball as in weighing ten. The proportion which the error bears to the total quantity weighed is, however, diminished when this quantity is increased. The use of a large number of bicycle balls for the determination of specific gravity in Experiments 3

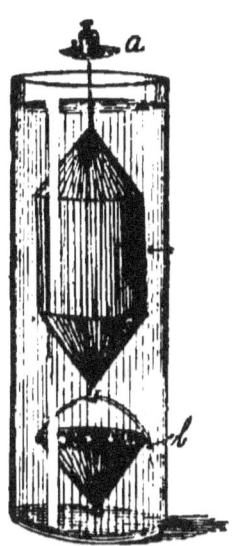

Fig. 9.

and 4 is a good example of the accuracy gained by the method of multiplication (§ 39).

¶ 12. **Corrections Disregarded in Experiment 4.** — In the last experiment we disregarded the effects of the buoyancy of air on the steel balls and on the brass weights, because these effects were so nearly equal, both being in air. Here, however, the balls are in water and the weights in air.

There is, therefore, nothing to compensate for the buoyancy of air on the brass weights. It is seen by reference to § 65 that 7 grams of brass are buoyed up by the air with a force of about 1 milligram; and as a Nicholson's hydrometer can float only about 4 times 7, or 28 grams, the effect of buoyancy on the weights cannot be greater than 4 milligrams. This error may generally be disregarded in comparison with the errors of observation. The manner of applying a correction for the buoyancy of air is explained in Experiments 8 and 9, also in §§ 65-68.

In calculating apparent specific gravity, no corrections need be taken into account; but the result should be expressed as the apparent specific gravity of a given body at a given temperature referred to water at a given temperature. The result will be affected somewhat by the density of the air, but hardly to a perceptible extent. The student is advised, as a matter of habit simply, to note the conditions of the atmosphere in which his weighings are performed (see Experiment 5).

EXPERIMENT V.

ATMOSPHERIC DENSITY.

¶ 13. **Determination of Barometric Pressure.** — The three conditions of the atmosphere which affect the results of physical measurement are barometric pressure, temperature, and humidity. Let us first consider how barometric pressure is observed. A very rough but serviceable form of mercurial barometer consists simply of a glass-tube ($a\ b$, Fig. 10), which, having been filled with mercury,[1] is inverted in a cistern of mercury (b). The mercury sinks in the closed end of the tube to a level a, above which there will be a nearly perfect vacuum.[2] As there is no pressure at a, to counteract the atmospheric pressure below, the mercury stands in the tube at a level (a) above the level (b) in the cistern. It is found by experiment[3] that the atmospheric pressure is transmitted through the cistern of mercury and the open end of the tube to a point, b, on a level with the surface of the mercury in the cistern. The atmospheric pressure is accordingly determined by

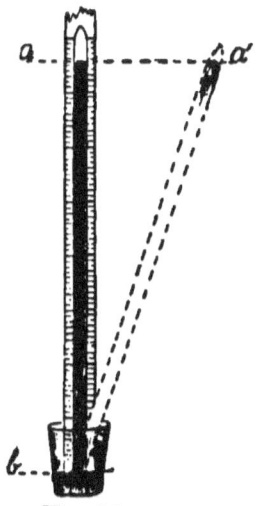

FIG. 10.

---

[1] The tube and the mercury must be perfectly clean and dry. For cleaning mercury, see Pickering's Physical Manipulation, I. 9.
[2] The "Torricellian vacuum." [3] See § 62.

the length, $a\,b$, of the column of mercury which it sustains.[1] The distance $a\,b$ can be measured by means of a graduated wooden rod, by which the tube is supported in a vertical position. The level b is first sighted in the ordinary manner with care to avoid parallax (§ 25); and the reading thus found is subtracted from that of the level a, obtained in a similar manner (see § 32).

In the case of a standard mercurial barometer the lower end of the column of mercury should always be looked at first, else a considerable error is likely to arise (§ 32); for even when the barometer ends in a large cistern of mercury, the level in this cistern must vary somewhat as more or less mercury rises into the tube. In some barometers this rise and fall is compensated by turning a screw (d, Fig. 11).

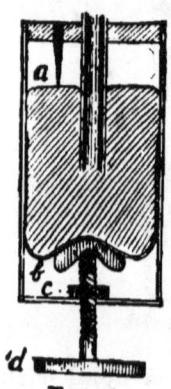

Fig. 11.

This raises or lowers the mercury in the cistern, and when a certain steel or ivory point (a, Fig. 11), just touches its own reflection, the level of the mercury is known to be at the right height. When the lower end of the mercurial column in the tube has been thus adjusted, the height of the upper end is usually read by a movable sight, provided with a vernier (§ 40). The lower edge of the sight is to be set on a level with the highest part of the mercurial column, so as to appear to be tangent to the meniscus or curved surface of the mercury (Fig. 12, a). To avoid par-

[1] See § 68.

¶ 14.] CORRECTIONS OF A BAROMETER. 19

allax (§ 25), a double sight is frequently used, consisting of two edges in the same horizontal plane, one in front of, the other behind the mercurial column. The student should find by direct measurement whether the distance from the zero-point (*a*, Fig. 11), to the lower edge of the sight (*a*, Fig. 12) is indicated correctly upon the scale of the barometer. If the reading of the barometer is in inches, it may be reduced to centimetres conveniently by Table 16.

FIG. 12.

Aneroid barometers are generally constructed so as to agree very closely with mercurial barometers. They will be found accurate enough for correcting the results of most physical measurements. If an Aneroid barometer is to be used, the student should compare its indication with that of a mercurial barometer, determined as explained above.

¶ 14. **Corrections of a Barometer.** — A small quantity of air almost always finds its way sooner or later into the space above the mercury in a barometer (*a*, Fig. 10), where it causes a slight depression of the column. To test a barometer for air, we tilt the tube *a b* (Fig. 10) into a new position *a' b*, being careful to keep the mercury in the cistern at a constant level, *b*, either by raising the cistern or by adding more mercury to compensate for that which flows into the tube. In the absence of air, the mercury should follow the horizontal line *a a'*, and should completely fill the tube when the inclination is sufficiently increased.

A simple way of correcting for air in a barometer is to adjust the angle $a'\,b\,a$ (Fig. 10) by trial, so that the space above a' is half that above a. By thus reducing the air to half its original volume, the pressure will be doubled;[1] hence a' will be as much below a as a is below its proper level. By measuring the difference between the levels, a and a', we find accordingly the correction for air. A correction of 2 or 3 *mm.* may be disregarded, as it will probably be offset by other corrections which the accuracy of the instrument will not justify us in considering. In case the correction is much larger than this, the barometer should be refilled with mercury. The filling of a standard barometer should be attempted only by a skilled workman. Unless perfectly free from air, such a barometer is little better than the rough instrument shown in Fig. 10.

In all exact readings of a barometer, the three following corrections are usually applied: (*a*) for expansion, (*b*) for capillary depression, and (*c*) for the pressure of mercurial vapor.[2] The temperature of the mercury in a barometer is found by a thermometer beside it. Let t be this temperature, reduced if necessary to the Centigrade scale (see Table 39), and let h be the height in centimetres of the mercurial column; then the correction for expansion is $.00018\,ht$, which is to be subtracted from the observed height. The object of this correction is to find

---

[1] This follows from the law of Boyle and Mariotte (§ 79).

[2] The reduction of a barometric reading "to the sea level" is not required for the purposes of physical measurement.

how high the mercury would stand if its temperature were 0° Centigrade. Since 1 $cm.$ of mercury when heated 1° Centigrade expands by the amount .00018 $cm.$ (see Table 11), h $cm.$ would expand h times as much; and h $cm.$ heated $t°$ would expand ht times as much, whence we obtain the correction in question. At the ordinary temperature of a room (20°), and at the barometric pressure, 75 $cm.$, this correction for expansion would be .00018 $\times$ 20 $\times$ 75 $cm.$ = 2.7 $mm.$ It is therefore useless to read a barometer (as is often done) to tenths or hundredths of a millimetre, when no correction for temperature is made. The correction given above may be applied to barometers with wooden or glass scales, the expansion of which may be neglected. When, however, the body of the instrument consists of steel, the coefficient .00017 should be used instead of .00018; and if the barometer is mounted in brass or white metal, the factor .00016 will be still more accurate. These numbers represent the difference of expansion between the mercury and the scale by which it is measured. For more accurate values see Table 18 a.

When the tube of a barometer is less than a centimetre in diameter, there is found to be a perceptible depression of the mercurial column due to " capillarity," or " surface tension," the general nature of which will be investigated farther in Experiment 67. The internal diameter of the tube should be found if possible by measuring a plug which fits it in the part where the column of mercury ends (see a, Fig. 10). A different method of calibration will be considered in Experiment 26. When the internal diameter is known, the correction for capillarity may be found roughly from Table 18 *b*. Thus for a tube 5 *mm*. in diameter, in which the height of the mercury meniscus is unknown, the capillary depression may be taken as 1.5 *mm*. In various barometers which are constructed so that the internal diameter cannot be measured, we generally assume that the instrument-maker has allowed for capillarity in adjusting his scale, and we therefore neglect this correction. It is customary, also, to neglect the effect of capillary phenomena in the cistern of mercury.

Owing to the evaporation of mercury into the space above it in the tube of the barometer, that space is never quite empty. The quantity of mercurial vapor which it contains is found to increase when the temperature increases, and also the pressure which it exerts. To allow for the slight depression of the mercurial column due to this cause, Table 18 *c* has been constructed from the results of actual observation. Thus for a temperature of 20°, we find that the mercurial column is depressed to the extent of 0.02 *mm*. by the pressure of its own vapor.

We have found in a particular case that 2.7 *mm*. should be subtracted from the observed height of a barometer on account of expansion; that 1.5 *mm*. should be added for capillarity and also 0.02 *mm*. to offset the pressure of mercurial vapor. The resulting correction is 1.18 *mm*., to be subtracted; or let us say, —1.2 *mm*. nearly. The student who employs a mercurial barometer should find in the same way an average correction for it. If an Aneroid is used, such a correction is found by comparing one reading at least with the corrected reading of a mercurial barometer. In the course of experiments which follow, readings of the barometer are needed only for slight corrections in the results of physical measurement. By applying to the barometer an average correction, much labor will be saved, and the error introduced will be insignificant.

¶ 15. **Determination of Atmospheric Temperature and Humidity.** — The temperature of the air of a room may be determined, with a sufficient degree of accuracy for most purposes, by an ordinary mercurial thermometer, the reading of which may be reduced from the Fahrenheit to the Centigrade scale by Table 39. The thermometer should be brought as near as may be practicable to the place where the temperature is required. It should, for instance, be inside of the balance case in very delicate weighings. It must not, however, be exposed to the rays of the sun, nor for any length of time to the heat radiated by a lamp or by the human body. When the greatest accuracy is desired, the bulb of the thermometer should be protected from radiation to or from surrounding objects, by a shield of polished metal.

The humidity of the atmosphere is most conveniently determined by a class of instruments of which the hygrodeik is an example. The indications of these instruments depend upon the cooling produced by evaporation (see § 88). It is found that when the bulb of a thermometer is covered with wet wicking (*a*, Fig. 13), its reading differs from that of an ordinary thermometer (*b*) by an amount depending upon the dryness of the air. When the air is completely saturated with moisture, as in a dense fog, there is no evaporation from the wet bulb, hence the two thermometers agree; if the air, however is heated, the fog disappears, evaporation begins, and the wet-bulb does not rise so high as the dry-bulb thermometer. On the other hand, when the air of the room is cooled sufficiently, either fog is formed or dew is precipitated on various objects; and the two thermometers again agree. The temperature at which this occurs is called the dew-point, and is calculated from the readings of the wet and dry-bulb thermometers by reference to Table 15, or by a special mechanical device, for the operation of which directions are usually furnished by the instrument-maker.

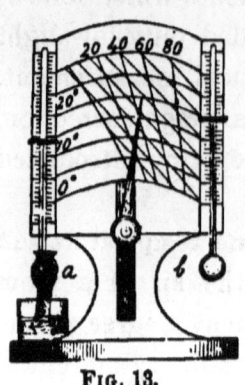

Fig. 13.

¶ 16. **Observation of the Dew-point.** — Unless a hygrodeik is known to give accurate indications, the latter should be confirmed by a direct determination of the dew-point, as follows: a polished metallic vessel is partly filled with water, and as much ice and salt are added as may be necessary to make a film of moisture condense on the surface. The temperature at which this first occurs is just below the dew-point. Soon, however, the contents of the vessel become warmer through contact with the air, and the film begins to disappear. The temperature is now a little above the dew-point. By observing carefully a thermometer with which the cold contents of the vessel are continually stirred, the dew-point may be determined within two limits, differing by less than one degree.

Care must be taken not to breathe on the metallic vessel, since the breath is much damper than the air of the room; and as there is more or less evaporation from all parts of the human body, even the hand should be kept as far away as possible.

¶ 17. **Relation of Relative Humidity to Dew-point.** The actual amount of moisture in a given quantity of air has been determined by extracting it through the action of certain hygroscopic substances, such as chloride of calcium, and measuring the gain in their weight. It is found that hot air can hold more moisture without forming fog than cold air. We have a common instance in the air of a room which, though apparently dry while warm, deposits moisture upon the window-panes by which it is cooled.[1] The ratio of the amount of moisture actually held in the air (at a given temperature) to the maximum amount (which can be held at that temperature) is called the relative humidity of the air. The relations between temperature, dew-point, and relative humidity do not follow any simple law; but if any two of these quantities are given, the third may be found by

[1] For a further illustration see list of Experiments in Elementary Physics, published by Harvard University, Exercise 22.

referring to Table 15, containing the results of various experiments.

It may be noted that the dew-point depends solely upon the amount of moisture in the air; that dry air has a lower dew-point and less relative humidity than moist air at the same temperature, while for a given dew-point the relative humidity increases with a fall of temperature, until fog is finally formed, or decreases as it becomes warmer until the air is practically dry. It should also be noted that dry air is denser than moist air. We must regard the latter as a mixture of air, not with water, but with steam, which is only about two-thirds as heavy as air. Hence in Table 20 the correction for moisture is negative.

¶ 18. **Determination of Atmospheric Density by means of a Barodeik.** — From the temperature, pressure, and humidity of the atmosphere, the determination of which has been explained above, the density of air may be calculated by the data of Tables 19 and 20. Whenever great accuracy is desired this calculation must be performed. For most purposes, however, the density of the atmosphere may be found from a single observation of a barodcik (Fig. 14), the principle of which is spoken

Fig. 14.

of in § 71. It is important to compare the indication of the instrument in at least one case with the calculated density of the atmosphere. A reading of the barodeik should accompany every weighing in which more than three figures are to be preserved, except when the pressure, temperature, and dew-point have been determined.

EXPERIMENT VI.

TESTING A BALANCE.

¶ 19. **Manipulation of a Balance.** — The delicacy of a balance depends upon the sharpness of the knife-edges (a and c, Fig. 15) from which the pans are suspended, also upon the sharpness of the central knife-edge (b) upon which the beam ($a\,c$) turns. In order that these edges may not become dull, the pans should be supported by some mechanical device at all times except when an observation is actually being taken. It is particularly important that they should be so supported when they are being loaded or unloaded, or when the balance is liable to be jarred in any other manner. In an ordinary prescription balance (Fig. 15), the pans rest upon the bottom of the case when the instrument is not in use. Such a balance is thrown into operation by turning a milled head outside of the case. The beam is thus raised as slowly as possible, so as not to injure the knife-edges by suddenly throwing weight upon them. It is not necessary in every case to raise the beam as far as it will go. As soon as the pointer moves decidedly to one side or the other, the beam should be slowly lowered again. In other cases a prolonged observation of the pointer must be made in order to decide in which direction the beam tends to incline. During such observations the beam should be raised to its fullest extent. Whenever accuracy is desired, the

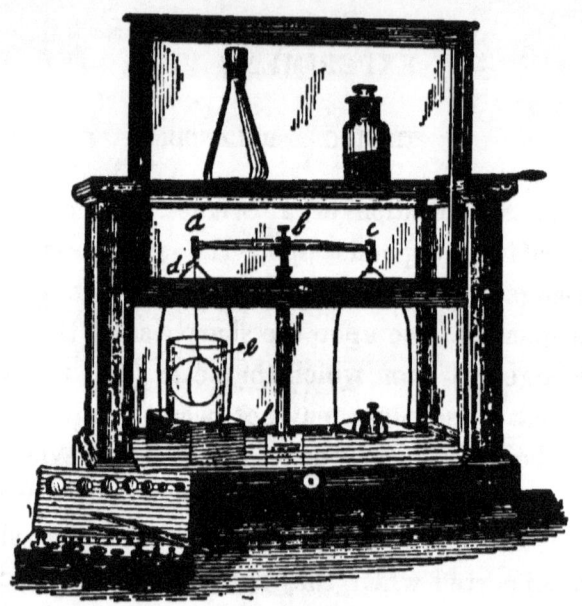

FIG. 15.

door of the balance case should be closed, in order to cut off currents of air; in fact, the door should never be opened except when the purposes of manipulation actually require it. This precaution is necessary to protect the instrument from moisture and dust, and is especially important when the air within the balance case is kept artificially dry by chloride of calcium or other hygroscopic material. The glass case should be cleaned when necessary with a damp cloth, to avoid charging it with electricity.[1]

Before weighing with a balance the case should be levelled and firmly supported, the scale-pans should be scrupulously cleaned and returned to their places, and any dust which may have collected on the knife-edges or their bearings should be cautiously removed with a camel's-hair brush. The beam is now thrown into operation by the mechanism already alluded to. If the instrument is correctly adjusted, the pointer attached to the under side of the beam will oscillate slowly and for some time through nearly equal arcs on either side of the central division of a scale (f, Fig. 15) directly behind it. If it tends to one side, that side is the lighter; and bits of paper or tinfoil should be fastened to the scale-pan until an exact balance is established.[2]

In loading the pans, pincers should be used as much as possible. In the case of the smaller weights, especially, contact with the fingers should be avoided. It makes no difference, theoretically, where the loads in the pans are placed; but many practical difficulties will be avoided by keeping them as nearly as possible in the centre. Both loads should be at the same

---

[1] By rubbing the glass at one side of a balance case with a piece of silk, a considerable error may be introduced into a weighing. The student should be cautioned, in general, against the effect of charges of electricity on delicate instruments. An eye-glass rubbed on the sleeve has been known to cause serious errors in physical measurement.

[2] See, however, first footnote, ¶ 26.

temperature as the air within the balance case; for though heat weighs nothing, a hot body may be lifted slightly by upward currents of hot air around it. With non-metallic loads we should avoid friction, which, as we have seen, may generate charges of electricity. When magnetic matter (as iron or steel) is to be weighed, all magnets (§ 126) should be removed from the immediate neighborhood. In an actual weighing, the scale-pans should be prevented from swinging, both on account of currents of air and because of the irregular motion given to the pointer.

¶ 20. **Method of Weighing by Oscillations.** — The reading of a pointer is usually taken while it is in motion, since much time would be lost in waiting for it to come to rest, and even then friction might stop it somewhat on one side of its true position of equilibrium. While in motion the pointer swings first to one side of its position of equilibrium, then to the other. The furthest point reached in a given swing to the right or to the left is called as the case may be a right-hand or a left-hand turning-point. Owing to friction, each swing is smaller than the one before it; hence the position of equilibrium is not exactly midway between any two successive turning-points. To avoid errors from this source we adopt the following rule: *observe any* ODD[1] *number of consecutive turning-*

---

[1] The object of making an odd number of observations is that the first and last may be on the same side; for in this case the turning-points on one side are on the whole neither earlier nor later than on the other side, and the gradual diminution of the swing affects each average alike.

points; find the average of those on the right and the average of those on the left; add these averages algebraically and divide by 2. The result is the point about which the oscillation is taking place, and at which the index tends eventually to come to rest.

It is convenient for many reasons to call the middle scale-division number 10, not 0, since otherwise plus and minus signs must be employed. In practice it is sufficient to observe three consecutive turning-points of the index.

It is frequently impossible to balance a given load exactly by any combination of weights which we are able to obtain. Let us suppose that with a weight, w, the index tends to rest at a distance from the middle-point equal to x scale-divisions; while with the smallest possible addition of weight, a, it tends to rest on the other side of the middle-point and at a distance from it equal to y scale divisions. Then the exact weight indicated for the load, l, is (see § 41),

$$l = w + \frac{ax}{x+y}.$$

The quantity $x + y$ is called *the sensitiveness of the balance to the weight* (a) *under the load* (l); and as it occurs in all exact estimations of weight by interpolation, it may be made properly the subject of further investigation.

¶ 21. **Determination of the Sensitiveness of a Balance.** — To test the sensitiveness of a balance with the pans empty, after carefully adjusting it as suggested in ¶ 19, we add a small weight, let us say 2 *cg*.

to the left hand pan. Instead of swinging about the middle scale-division, which we have agreed to call number 10, it will swing about a new point corresponding, let us say, to number 12·6 on the scale. This would show that the balance is sensitive to the extent of 12·6 — 10, or 2·6 divisions for 2 *cg.*, or 1·3 divisions per *cg.*, when the pans contain little or no load besides their own weight. This fact is recorded by making a cross (as in Fig. 16) on a piece of co-ordinate paper at the right of the number 0, representing the load, and below the number (1·3) representing the sensitiveness in question.

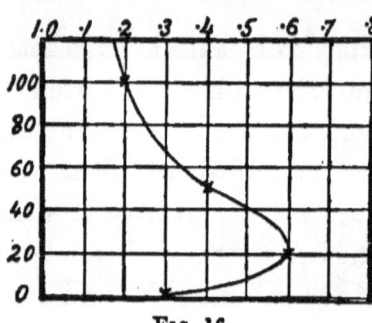

Fig. 16.

We now place, let us say, 20 grams in each pan, and find as before the sensitiveness per centigram. It will not necessarily be the same as when the pans are empty; in fact, a difference is almost always observed.[1] The sensitiveness is then found with 50 grams in each pan, and finally with 100 grams in each pan. Thus, in an actual case, a balance which was sensitive with the pans empty to the extent of 1·3 divisions per *cg.*, was affected to the extent of 1·6 divisions per *cg.* with 20 *g.* in each pan, 1·4 divisions

[1] It will be shown in ¶ 22 that the effect of a load on the sensitiveness of a balance cannot be anticipated; hence the student who records faithfully what he sees, not what he expects to see, will here as elsewhere in Physical Measurement, be likely to obtain the most accurate results. (See § 30.)

per *cg.* with 50 *g.* in each pan, and 1·2 divisions per *cg.* with 100 *g.* in each pan. These results are recorded, as before, by crosses in the proper places (see Fig. 16), and a curve is drawn by a bent ruler through these crosses. This curve enables us to find approximately the sensitiveness of the balance under any ordinary load by the method explained in § 59.

When we know the sensitiveness (s) of a balance to 1 *cg.*, a single observation of the pointer is sufficient to determine exactly the weight indicated. If w is the lighter weight (in the pan toward which the pointer inclines) and x the number of scale-divisions between the resting point of the index and the middle of the scale, the load (l) indicated is found by substituting s for $x + y$ and .01 for a in the formula of ¶ 20; or

$$l = w + \frac{.01\, x}{s}.$$

¶ 22. **Conditions on which the Sensitiveness of a Balance Depends.** — In order that a balance may move perceptibly under the influence of a very small weight added to either pan, the central knife-edge (b, Fig. 15) on which the beam turns must not only be sharp (¶ 19), but must pass nearly through the centre of gravity. If the centre of gravity is above this knife-edge, the balance will be "top heavy." This difficulty must be remedied by attaching a bit of sealing-wax to the pointer below the knife-edge b, or by lowering the centre of gravity in any other manner.[1] If on the other hand the centre of gravity is too low, the balance will be too steady, and it will not respond sufficiently to a small change in the load. In this case it is necessary to fasten a small weight to the balance beam, somewhere above the knife-edge b, or otherwise to raise its centre of gravity.

When the balance-pans are loaded, new considerations come in. Since in all positions of the beam the loads hang vertically beneath their respective knife-edges, the result is the same as if they were concentrated at those knife-edges. Let us suppose that the instrument has been adjusted so as to be sufficiently sensitive when the pans are empty. In order that it may remain equally sensitive when loaded, the three knife-edges must be in the same straight line, as in A, Fig. 17. If the two outer knife-edges which

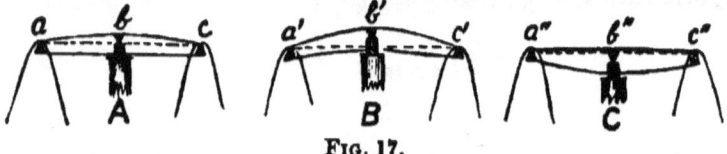

FIG. 17.

bear the loads (see a'', c'' in C) are distinctly above the central knife-edge (b''), the combined effect of the loads will be towards unstable equilibrium; or if the outer knife-edges (see a', c' in B), are below the central knife-edge (b'), the combined effect of the loads will be to steady the balance, and hence to diminish its sensitiveness. There are therefore three types to which a balance beam may belong, repre-

---

[1] A movable screw or counterpoise is provided in some balances for the purpose of raising or lowering the centre of gravity.

sented by the three diagrams, A, B, and C. In the first, the load does not affect the sensitiveness, except in so far as friction may be concerned; in the second, it lessens it; in the third, it may increase the sensitiveness until the balance actually becomes "top heavy."

A common balance may belong successively to all three of the types, C, A, and B. Let us suppose that with the pans empty the extremities of the beam are bent upward, as in C. With a medium load, the beam may be straightened, as in A, and with a still greater load the ends may be bent downward, as in B.

Such a balance would be more sensitive with a small load in each pan than when the pans were empty; because a small load, being insufficient to straighten the beam, would raise its centre of gravity[1] as in C; but when already heavily loaded, so that the beam is bent downward as in B, the further addition of weight would lessen its sensitiveness. The curious shape of the curve found in the last section (Fig. 16), is thus accounted for.

¶ 23. **Determination of the Ratio of the Arms of a Balance.** — The balance is now readjusted if necessary as in ¶ 19, so that the pointer swings accurately about the central division of the scale when the pans are empty, and the 100 gram weight is balanced against its equivalent as before, only that small weights are added to one side or to the other to

---

[1] A balance, though stable with a heavy or with a medium load, as well as when the pans are empty, may actually become "top heavy," with a small load in each pan. In such a case, the centre of gravity should be permanently lowered.

bring the pointer as nearly as possible to the central division, and the exact weight estimated as in ¶ 21, considering as the load, l, that weight which is apparently the larger. The loads in the two pans are now interchanged, readjusted by the use of the small weights, and compared exactly as before. The pans being once more emptied, the pointer should swing about the central division, otherwise the balance must be readjusted and the process described in this section must be repeated until the equilibrium of the balance remains undisturbed.

The object of testing the balance, as above, with equal weights in the opposite scale-pans, is to discover any inequality which may exist in the length of the balance arms ($a\,b$ and $b\,c$, Fig. 17). Such an inequality might seriously affect the accuracy of results, and we have no right to neglect it even in ordinary weighings without some test similar to the one described. It is true that by the method of double weighing (see § 44), errors due to the inequality of the balance arms may be eliminated; but double weighings are sometimes impracticable, as in the case of a body of variable weight, or in a very long series of determinations. In such cases the inequality of the balance arms should be found by a careful and extended series of observations. For the purposes of this course of experiments, a single determination will suffice. The ratio of the balance arms is calculated therefrom as explained in the next section.

¶ 24. **Calculation of the Ratio of the Balance Arms.** — If the arms of a balance are unequal, it is important to know from which arm the unknown weight is suspended. To avoid the necessity of mentioning in each case the pan containing the load in question, it is customary to place the unknown weight at the left hand whenever a single weighing is to be made. In this way the known weight, consisting generally of several small pieces, is conveniently adjusted by the right hand.

To find the proportion which the weight on the left arm always bears to the weight on the right arm, we need only a single comparison between two known weights. As these weights are inversely as their respective arms (see § 113), the proportion in question is equal to the ratio of the right arm to the left arm. Thus if (in an extreme case) 101 grams in the left-hand pan balance 100 grams in the right-hand pan, the right arm must be $\frac{101}{100}$ or 1.01 times as long as the left arm. All weights in the left-hand pan are therefore 1% greater than those which balance them in the right-hand pan; hence to find the value of an unknown weight in the left-hand pan we multiply that of the known weight in the right-hand pan by 1.01. The ratio of the balance arms is in general that number by which the known weight must be multiplied in order to find the unknown weight which balances it. We usually require, as we have seen, the ratio of the right arm to the left arm. This is found by dividing a known weight in the left-hand pan by a known weight in the right-hand pan which balances it.

The object of interchanging the two weights in

¶ 23, each nominally equal to 100 grams, is to avoid mistakes arising from a difference between the two weights in question. If no such difference exists, the interchange will not affect the result. Otherwise to find the ratio of the balance arms, we take the average of the two weights in the left-hand pan, and divide it by the average of the two weights in the right-hand pan. In taking these averages we accept the nominal values of the weights in question, any errors in which are practically eliminated by the method of interchange (§ 44) here adopted.

EXPERIMENT VII.

CORRECTION OF WEIGHTS.

¶ 25. **Process of Testing a Set of Weights.** — The brass 1 gram weight is first balanced against all the smaller weights, which should together be equal to 1 gram; then each 2 gram weight against the 1 gram plus the smaller weights; then the 5 gram weight against the two 2 gram weights plus the 1 gram; then in the same way the 10, 20, 50, and 100 gram weights, each against its equivalent. Whenever there are two ways of making an equivalent, that selection is made by which the fewest weights may be employed. (See § 36, 2d ed.) The 100 gram weight is finally balanced against a standard.[1] In

[1] The standard should be of the same material as the set of weights employed, that is, of brass; but if any other material is used, a correction must be made for the unequal buoyancy of the atmosphere upon the loads in the two pans. See § 67 and Table 21.

each case, where two weights are balanced, the difference between them is estimated by the method of vibration (¶ 20), and recorded as will be explained below. To avoid corrections named in the last experiment, the method of double weighing is used in every case.

¶ 26. **Estimation of Tenths in Weighing.** — In a long series of weighings, as in testing a set of weights, it is hardly thought to be advisable (see, however, § 33) to record each turning-point of the index as in ¶ 20. The student who wishes to make any extended use of the balance should learn to estimate correctly the point of the scale about which the index is swinging, and hence the number of divisions from the middle of the scale [1] to the point where the index tends to rest; to carry this number in the head while finding by inspection of figure 16 (see ¶ 21 and § 59) the sensitiveness of the balance under the load in question,[2] and to divide mentally the number thus carried in the head by that representing the sensitiveness of the balance, or the effect of 1 cg. (See general rules for interpolation, § 41.) He will thus find the fraction of a centigram necessary to make the index swing about the middle-point of the scale, and will

---

[1] Instead of adjusting the balance as in ¶ 19, so that the index may swing about the middle-point of the scale, the advanced student may often prefer to observe accurately the point about which the index actually oscillates when the pans are empty, and *to measure all distances from this point.*

[2] It is sometimes quicker to add one centigram to the lighter pan, and thus to re-determine the sensitiveness. In many cases the sensitiveness may be recalled from memory with a sufficient degree of exactness.

record the number of milligrams nearest to that fraction with the proper algebraic sign.

Thus if with a weight marked 10 g_1 in the left-hand pan and with 10 g_2 in the right-hand pan, the index swings about a point corresponding to 10·3 of the scale,—that is, 0·3 divisions to the right of the middle-point,—and if the sensitiveness of the balance with a load of 10 grams is about 1·5 divisions per centigram (see Fig. 16, ¶ 21), the weight 10 g_1 is clearly heavier than 10 g_2 by $0·3 \div 1·5 = \frac{1}{5}$ cg. or 2 mgr. We record such an observation as follows:

$$10 \, g_1 = 10 \, g_2 + 2 \, mgr.$$

In the same way we enter the result of placing 10 g_1 in the right-hand pan and 10 g_2 in the left-hand pan; and if there is any difference, we find the average excess of 10 g_1 over 10 g_2, or the reverse.

¶ 27. **Calculation of the Corrections for a Set of Weights.**—Any one familiar with algebra can find the relations existing between the different weights of a set from a series of equations obtained as in the last section. The following suggestions may however be useful. Call the value of the 1 gram weight G; find the total value of the smaller weights (100 cg.) in terms of this. For instance, let

$$100 \, cg. = G + 1 \, mgr.$$

Then find the value of the 2 gram weights, 2 g_1 and 2 g_2 in terms of G. If for example,

$$2 \, g_1 = 100 \, cg. + G - 1 \, mgr.,$$

we find, substituting for 100 cg. its value, $G + 1$ mgr.,

$$2 \, g_1 = G + 1 \, mgr. + G - 1 \, mgr. = 2 \, G;$$

CORRECTION OF WEIGHTS.

and if still further, it has been observed that
$$2 g_2 = 2 g_1 + 2\ mgr.,$$
we find similarly
$$2 g_2 = 2\ G + 2\ mgr.$$
Again, if by observation
$$5 g = 2 g_1 + 2 g_2 + G + 1\ mgr.,$$
we have
$$5 g = 2\ G + 2\ G + 2\ mgr. + G + 1\ mgr.$$
$$= 5\ G + 3\ mgr.$$

In the same way we find the values of all the weights in terms of G, until we come finally to the standard. Knowing the standard in terms of G, we find G in terms of the standard. The corrected value of G should be expressed in grams and carried out to five places of decimals. Substituting this value in all the equations, we obtain finally the correction in *mgr.* for each weight belonging to the set from 1 gram upwards.

This method of framing and reducing equations is not peculiar to a set of weights. The student may substitute for it, if he prefers, the correction of a set of standard electrical resistances, which he will learn how to compare in Experiment 87. The same method may be applied to any other standards capable of being arranged like a set of weights, so that each one may be compared with an equivalent made up of the others below it. The general principle by which such a standard set is corrected is one of the best illustrations of the method of multiplication (§ 39) upon which nearly all measurements are founded.

EXPERIMENT VIII.

WEIGHING WITH A BALANCE.

¶ 28. **Determination of Weight in Air by a Balance.**
— The apparent weight of a body in air may be found approximately, as has been explained in Experiment 1, by placing it in one pan of a balance — the left being understood unless otherwise stated (see ¶ 24) — and finding by trial (¶ 2) the requisite number of weights to counterpoise it. The accurate determination of weight in air differs from this rough method chiefly in the delicacy of the instrument employed, and in the consequent care of manipulation (see ¶ 19). In this, as in all other accurate determinations with the balance, unless otherwise stated, it is assumed that the method of weighing by oscillations is employed (¶ 20).

The object recommended for this experiment is a glass ball, the weight of which will be needed later on in the course. To prevent it from rolling out of the pan, it may be set in the middle of a small ring of known weight, which we will suppose to be counterpoised with one of equal weight in the opposite pan.

It is necessary in this experiment either to know the ratio of the balance arms (see ¶ 23), or to employ the method of double weighing (§ 44) as in Experiment 7. The density of air must also be determined by an observation of the barodeik (¶ 18), or by an observation of the atmospheric pressure, temperature, and humidity (¶¶ 13-15). We must also know the material, and hence approximately the densities of both the object weighed and the weights with which it is counterpoised. These densities may be found with a sufficient degree of accuracy by referring to Tables 8-11. The correction of apparent weights to *vacuo* is then made as explained in § 68.

EXPERIMENT IX.

THE HYDROSTATIC BALANCE, I.

¶ 29. **Determination of the Density of Solids by the Hydrostatic Balance.** — An arch is placed over a balance pan as in Fig. 18, so as not to interfere with its free vibration; and on the middle of the arch is set a beaker. The glass ball weighed in the last experiment is now bound in a network of fine wire and suspended by a single strand from the hook of the balance, so as to clear the bottom of the beaker. The latter, being moved if necessary so that its sides may not touch the ball, is filled with a quantity of distilled water sufficient to cover,[1] in all positions of the balance, both the ball and its network of wire. All bubbles of air

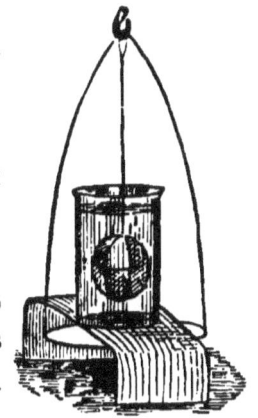

FIG. 18.

[1] A small loop of wire, projecting above the surface, may completely ruin a determination.

clinging to the ball, or wire, must now be removed with a camel's-hair brush. The suspending wire, being likely to attract grease or other foreign matter which repels water, is cleaned if necessary, so that it may be kept wet for a distance of about one centimetre above the level of the water, by the continual oscillation of the balance. The capillary phenomena already noticed in ¶ 5 are thus reduced to a small and nearly constant amount.[1]

By these adaptations the instrument which we employ has been completely transformed into a "hydrostatic balance," by which the weight of the ball and wire in water may now be found, as in the last experiment, by counterpoising it with weights in air (see Fig. 15, ¶ 19). The method of weighing by oscillations is not, however, recommended in the case of a hydrostatic balance; but rather a direct observation of the pointer in its position of equilibrium, which, owing to fluid friction, is quickly reached.

Apart from friction, the sensitiveness of a hydrostatic balance is always somewhat less than that of the same balance when used for measuring weights in air,[2] and must therefore be re-determined by adding a centigram to the smaller of the two loads when nearly balanced and observing the result (see ¶ 21). In this, as in all experiments with the hydrostatic

---

[1] The use of spirits of wine to diminish still further the capillary action (Trowbridge, "New Physics," page 17), is not recommended to beginners, on account of the danger of its mixing with the water and thus affecting its density.

[2] The variable amount of water displaced by the suspending wire tends to increase the stability of the balance.

balance, the temperature of the liquid should be observed both before and immediately after finding the weight of a solid in it.

The weight of the wire in water must be found separately in the same manner and under the same conditions as before.[1] The ball is removed from the network of wire so as to leave the latter undisturbed in so far as possible, and water is added to the beaker in order that the same amount of wire may be submerged in each case. It may even be necessary, if a coarse wire is used, to adjust the level of the water exactly to a given mark, and if the network is bulky, to raise or lower the temperature of the water to the same point as before.

The apparent weight of the ball in water is found by subtraction, and reduced to *vacuo* by the principle of § 67. The difference between the apparent weights in air and in water gives the apparent weight of water displaced (§ 66), and hence the volume displaced (see Table 22). The difference between the weight of the ball *in vacuo* (¶ 28) and its weight in water (reduced to *vacuo* as explained above) gives, by a strict interpretation of the Principle of Archimedes (§ 64), the weight *in vacuo* of water displaced, and hence also its volume (by Table 23). We have thus two methods of calculating volume, of which the first is more generally useful, as it does not require any previous reduction of weights to *vacuo;* but the

---

[1] Precautions similar to those which follow are necessary whenever a method of difference is employed. For further illustration see § 32.

second is more rigorous, because, depending upon weights *in vacuo*, the results will not be affected by variations of apparent weight due to changes in atmospheric density. The latter should therefore be employed when any considerable time elapses between the determinations of weight in air and in water. The density (or average density) of the ball is finally calculated (see ¶ 1) by dividing its weight *in vacuo* by its volume. (See ¶ 4, also § 68.)

EXPERIMENT X.

THE HYDROSTATIC BALANCE, II.

¶ 30. **Determination of the Density of Liquids by the Hydrostatic Balance.** — The experiment consists essentially of a repetition of Experiment 9, substituting, however, for distilled water some other liquid of greater or less buoyancy.

Various modifications of this experiment may be necessary according to the nature of the liquid used; for instance in the case of strong acids, platinum wire must be substituted for iron, which would be speedily dissolved, and even platinum cannot be used in *aqua regia*. To avoid fumes in the balance case, the suspending wire is sometimes carried down through a series of small holes to a beaker below. To avoid evaporation, in the case of volatile liquids, the beaker should always be covered with cork or cardboard perforated for the suspending wire. The same precaution should be taken when moisture is likely to be absorbed. In some liquids scarcely any bubbles are formed; in others, such as glycerine, it may take hours to remove them, though their formation may be prevented if the glycerine is poured in a continuous stream down the sides of the beaker. In most liquids the effects of temperature are greater than in the case of water (see Table 11), hence the thermometer must be read with the greatest care. It is well to warm or cool the liquid (and hence also the ball) to the temperature of the water in Experiment 9, to avoid all corrections for temperature.

¶ 31. **Calculation of the Density of Liquids by the Hydrostatic Method.** — We find in the same way as in Experiment 9, the apparent weight of the ball in the liquid, allowing for the wire as before; and from this we subtract the weight of air displaced by the brass weights (see § 67), to find the true weight of the ball in the liquid. The difference between its true weight in the liquid and that *in vacuo*, already found (¶ 28), is equal to the weight *in vacuo* of the liquid displaced. This follows from the Principle of Archimedes (§ 64).

The volume of liquid displaced is of course equal to the volume of the ball, which will not differ perceptibly from the value previously determined (see end of ¶ 29) if the temperatures of the two experiments are nearly the same. If this is not the case, it is necessary to allow for an expansion or contraction of the glass, at the rate of about one part in 40,000 for every degree Centigrade. (See Table 8 *b* and § 83.)

The weight *in vacuo* of the liquid displaced is finally divided by its volume to find its density.

The weight *in vacuo* may be checked by calculating the apparent weight of the liquid displaced, as in Experiment 9, then reducing at once to weight *in vacuo* by applying the necessary factor from Table 21, as explained in § 68, using the density already calculated. This latter method is slightly inaccurate, as has been stated before (¶ 29), on account of its disregarding variations of atmospheric density during the course of experiments.

In determining the density of water by the hydrostatic balance, the weight displaced may be found as in Experiment 9 or 10; but the volume displaced cannot be calculated in the manner explained above, because the tables which we employ themselves depend upon the density of water. It is necessary to calculate the volume of the solid immersed from actual measurements of its dimensions[1] (see ¶ 1). By this method, essentially, with the aid of instruments of precision, accurate determinations of the density of water have been made (see Table 25). The student will have an opportunity in Experiment 19, to confirm these determinations within the limit of accuracy of the instruments which he employs.

---

[1] The volume, v, of the glass ball may be calculated from its diameter, d, by the formula, $v = \cdot 5236\, d^3$. In place of the glass ball we may use, for purposes of illustration, the rectangular block whose volume has already been determined in Experiment 1. If it floats in water, a lead sinker may be attached to it. The sinker must remain in place after the block is removed, in order that its weight may be allowed for. A spring balance may be used to find roughly the weight of water displaced. See Exercises 7–10 in the Descriptive list of Experiments in Physics published by Harvard University.

EXPERIMENT XI.

CAPACITY OF VESSELS.

¶ 32. **Determination of the Capacity of a Specific Gravity Bottle.** — Any bottle with a solid stopper of ground-glass may be used for finding the specific gravity of liquids; but when solids are to be introduced, one with a wide mouth will be needed. The capacity of the bottle is determined in the following manner. The bottle is first washed in perfectly pure water, then dried with a cloth inside and out, and afterwards still more thoroughly dried with a hot air-blast.[1] The weight of the bottle is found within a centigram, then the bottle is alternately dried and weighed until by the agreement of two successive weighings, the drying is known to be complete. The last weight found, if confirmed by the method of double weighing as in ¶ 28, is the apparent weight of the bottle in air. It is understood that the stopper is always weighed with the bottle. In this case, it should be placed in the scale-pan beside the bottle, so that the density of the air may be the same inside and out. The bottle, which will be warmed by the hot air-blast, must be allowed time to cool to the temperature of the room before the weighing is completed, since otherwise currents of hot air might seriously affect the result (see ¶ 19).

[1] When a hot air-blast cannot be had, the bottle may be dried by rinsing it out several times with a small quantity of alcohol, and exposing it for a few minutes to a draught of air.

The bottle is then filled with distilled water at an observed temperature, not far from that of the room; then closed in such a manner (see Fig. 19) as to allow all bubbles of air to escape.[1] The outside of the bottle is then carefully dried with a cloth or blotting-paper.

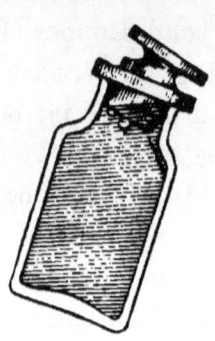

Fig. 19.

The weight is again found with the same degree of accuracy as before, and immediately afterward the temperature of the water and the density of the air (¶ 18).

The difference between the two apparent weights of the bottle containing air and water, respectively, is equal to the apparent weight in air of the water which it contains (§ 66); this weight of water multiplied by the space occupied (at the higher of the two observed temperatures, see ¶ 33) by a quantity of water weighing apparently 1 gram (in air of the observed density, see Table 22), gives the total space occupied by the water, or in other words the capacity of the bottle at the observed temperature.

¶ 33. **Effects of Varying Temperature on a Specific Gravity Bottle.** — It is hardly necessary, in the experiments which follow, to allow for the expansion of the glass bottle due to changes of temperature which

---

[1] If the shape of the stopper makes this impossible, it must be altered by grinding or by filling up any hollows in it with paraffine or other material not acted upon by ordinary liquids. In this case the weight in air must be re-determined.

it is likely to undergo.[1] In a laboratory, maintained as it should be at a nearly constant temperature, these changes will be slight. Unless, however, special precautions are taken to keep the water in the bottle at a constant temperature, serious errors are likely to arise. These errors will be still greater in the case of certain other liquids which we shall employ. The expansion of alcohol, for instance, will be found to be several hundred times as great as that of glass (see Table 11).

Let us first suppose that the liquid which fills a closed bottle is gradually cooling, and hence in the process of contraction. A bubble will soon be formed. This need not, however, give rise to apprehension if the initial temperature (at which the bottle was filled) has been correctly observed; for the weight of the liquid will not be changed by its contraction, and the bubble weighs practically nothing. We may therefore determine the weight of a liquid which fills a bottle at an observed temperature, after it has fallen below that temperature.

Now, let us suppose that the liquid is growing warmer; and hence, expanding, that it forces its way out by the stopper, yet clings to the bottle. Unless the liquid is volatile or hygroscopic,[2] its weight

---

[1] The capacity of a vessel increases by the same amount as the volume of a solid of the same material which would exactly fill the vessel. In the case of glass, this increase is at the rate of about 1 part in 40,000 per degree Centigrade.

[2] Hygroscopic liquids, such as sulphuric acid or chloride of calcium, should be slightly warmed before the experiment, so that they may be weighed while cooling.

will be unchanged, and hence may be determined at leisure. If, however, the liquid evaporates immediately (as many liquids do) on contact with the air, there will be a continual loss of weight. In such cases, we must find the temperature as nearly as possible at the time of weighing, when it will be seen that the quantity of liquid weighed exactly fills the bottle.

In practice, both the initial and final temperatures are usually observed; the former just before the insertion of the stopper, the latter immediately after completing the weighing. We notice that with a non-volatile liquid, the initial temperature is always required; and the same statement applies to a volatile liquid which is cooling; but with a volatile liquid in general it is the *maximum* temperature which we wish to determine. In no case do we take the mean of the two temperatures before and after the experiment.

The liquids which we employ should be warmed or cooled if necessary, so that they may be nearly at the same temperature as the room; since otherwise the rapid changes of temperature which must ensue (§ 89) would make an accurate observation of the thermometer impossible. Errors in weighing might also be introduced, owing to currents of hot or cold air (¶ 19). In the case of certain liquids (as ether) which are apt to become cold through evaporation,[1]

---

[1] Care must be taken in general to prevent evaporation; and especially in the case of impure liquids, the strength of which would be affected by the escape of the more volatile ingredients.

there is danger that moisture may be condensed on the sides of the containing vessel (see ¶ 17). Particular care must be taken in the case of water, when below the temperature of the room; lest through the humidity of the air or from other causes it should fail to evaporate as fast as it is driven out of the bottle. Any moisture collected around the stopper should be removed with blotting-paper before making a final adjustment of the weights.

EXPERIMENT XII.

DISPLACEMENT I.

¶ 34. **Determination of Displacement by the Specific Gravity Bottle.** — The experiment consists essentially of a repetition of Experiment 11, with a bottle already partly full of sand, or any other substance insoluble in water. The capacity of the bottle for water is evidently less than before by an amount exactly equal to the space which the sand takes up; hence the latter can be found by subtracting the new capacity from the old. This method of determining volume is especially convenient in the case of powders, which cannot easily be suspended from a hydrostatic balance.

Certain modifications of the methods used in Experiment 11 are introduced when finely divided substances are employed. Even with sand considerable difficulty may be found in removing the bubbles of air which cling to it under water. By continual shaking with water in a well-stoppered bottle, this air may generally be freed from the sand.[1] To obtain dry sand, it should be heated before the experiment to a temperature above 100°.

The same process may be used to dry various powders not easily melted or decomposed by heat; but others require special precautions belonging to the province of Chemistry rather than Physics.

It may be observed that the apparent weight of the solid used in this experiment is incidentally determined; for we have only to subtract from the apparent weight of the bottle with it that of the bottle without it as found in the last experiment. The density of the solid may therefore be calculated as in Experiment 9.

¶ 35. **Illustration of the Principle of Archimedes.** — To understand what is meant by the water displaced by a solid, the bottle may be filled with water as in Experiment 11, then the solid may be introduced; water will be literally displaced, and if the whole quantity thus driven out of the bottle could be collected and weighed, we should have a direct measurement of the water displaced by the solid. In practice we prefer to find this by difference.

If we call s the apparent weight of the sand, b that of the bottle, w that of the water which fills it, and d that of the water displaced by the sand, the weights observed are (1) b and (2) $b + w$ in Experiment 11,

---

[1] An air-pump greatly facilitates the process, but unless special precautions are taken the water is apt to bubble over into the receiver and to find its way into the valves of th pump.

(3) $b + s$ and (4) $b + s + w - d$ in Experiment 12. The apparent weight of water which fills the bottle is the difference between the first and second observations, or (2) — (1), but when the sand is already in the bottle the quantity of water required is the difference between the last two observations, or (4) — (3); hence the quantity displaced is [(2) — (1)] — [(4) — (3)].

Now the weight of the sand in air is evidently the difference between the first and third observations, or (3) — (1); its apparent weight in water is the difference between the second and fourth,[1] or (4) — (2); its loss of weight in water is therefore [(3) — (1)] — [(4) — (2)]. This is seen by comparison to be identical with the expression above for the weight of water displaced.

The student who finds difficulty in realizing how the apparent weight or loss of weight of a solid in water can be found by the specific gravity bottle may repeat these measurements with a hydrostatic balance, using a cup to hold the sand in place of the network of wire employed in Experiment 9 to hold the glass ball; or he may find the weight and loss of weight in water of the steel balls used in Experiment 4 by means of the specific gravity bottle. The Principle of Archimedes (§ 64) states that loss of weight

[1] In both observations we have the same weight of the bottle, and the same hydrostatic pressure of the water upon the bottom or sides of the bottle (§ 63); the only difference is the downward pressure of the sand, which is present in (4) and absent in (2). This pressure exerted under water is what we call the weight of the sand in water.

in water (which we think of as determined by hydrostatic methods) is equal to the weight of water displaced (which we think of as determined by a specific gravity bottle). The agreement of the results obtained by hydrostatic methods with those from the specific gravity bottle may serve therefore either as an illustration of this principle or as a mutual confirmation of these results.

EXPERIMENT XIII.

DISPLACEMENT II.

¶ 36. **Determination of the Volume and Density of Solids Soluble in Water.** When owing to the solubility in water of the substance employed, the method explained in the last experiment cannot be applied, it remains only to find some other fluid of known density in which that substance is insoluble. The various products of the distillation of petroleum are especially suited to this purpose, since they dissolve few (if any) ordinary substances which are soluble in water. We may occasionally, with great care, use a saturated aqueous solution of the substance whose density is to be determined, or a liquid which has been allowed to act chemically upon an "excess" of that substance, since in either of these cases the liquid will have no further action on the solid. Gases may also be employed; but on account of the difficulty of measuring their weight correctly even by the most delicate balances, it is customary to estimate the quantity present by a direct or indirect measurement of its volume.[1] Owing, however, to the tendency of certain substances to absorb large quantities of gas, all such methods may lead to erroneous and even absurd results.

For sake of simplicity we will choose the liquid whose density has been determined in Experiment 10, and for the solid some substance insoluble in that liquid; and in order that the density of the liquid may be the same as before, it should be warmed or cooled if necessary to the temperature observed in Experiment 10. With such a solid and liquid, Experiment 12 is to be essentially repeated.

¶ 37. **Calculation of Volume and Density by the Use of Specific Volumes.** We have already seen how the weight of water displaced by a solid may be found either by the hydrostatic balance (Experiment 9) or by the specific gravity bottle (Experiment 12). By the same methods we may obtain the weight of any other fluid displaced by a solid. We have already applied this principle in Experiment 10 for determining the density of a liquid. Knowing the weight in grams and the number of cubic centimetres displaced, we found by division the weight of 1 cu. cm. It would have been equally simple to interchange the divisor and dividend, and thus to find the space in cu. cm. occupied by 1 gram. This is sometimes called the *specific volume* of a liquid.

The mutual relations existing between the weight

---

[1] For a description of the "Volumenometer," see Trowbridge's New Physics, Experiment 31.

w, the volume v, the density d, and the specific volume s, of any substance are given by the equations

$$d = \frac{w}{v},\ s = \frac{v}{w},\ \therefore\ s = \frac{1}{d},\ v = w\,s,\ etc.$$

The specific volume is therefore technically the "reciprocal" of the density. To find it we divide unity by the density already determined in Experiment 10, or by that which we may find from Experiment 14.

We have already used specific volumes in Table 23 (see ¶ 29), and we know that the weight *in vacuo* of the liquid displaced, multiplied by its specific volume,[1] gives the actual volume displaced, which is of course equal to that of the solid causing the displacement. The volume of the solid enables us to reduce its apparent weight to *vacuo* (§ 67), and hence to calculate its density (§ 68).

EXPERIMENT XIV.

DENSITY OF LIQUIDS.

¶ 38. **Determination of the Density of a Liquid by the Specific Gravity Bottle.** We have already found the weight of a bottle containing water and air, and we have calculated its capacity; it remains only to find its weight when filled with any other fluid, in order

---

[1] The student should bear in mind that the specific volume here employed is the space occupied by a quantity of liquid weighing 1 gram *in vacuo*, not that which weighs apparently 1 gram in air. True specific volumes must be multiplied by true weights *in vacuo* to find actual volumes. Apparent specific volumes (see Table 22) are intended to give the same result with apparent weights in air.

that the density of that fluid may be determined. For the purpose of comparison we will choose the liquid already used in Experiments 10 and 13, and warm or cool it, as nearly as may be convenient, to the temperature of those experiments. The actual temperature should be observed for reasons explained in ¶ 33, both before and immediately after weighing. The barodeik should also be read, in order to make sure that no great change has taken place in the course of our experiments with the specific gravity bottle, since otherwise its apparent weight in air must be re-determined.

The apparent weight of a quantity of alcohol sufficient to fill the bottle is found by subtracting that of the bottle with air from that of the bottle filled with alcohol, and is reduced to *vacuo* as explained in § 67. The density is then calculated by dividing the weight *in vacuo* by the capacity of the bottle, from ¶ 32. The strength of the alcohol is finally found by reference to Table 27, using a process of double interpolation (see § 58). The strength of the alcohol may also be calculated from the data of Experiment 14; and even if the temperatures in Experiments 10 and 14 differ considerably, the two results should agree in respect to strength.

EXPERIMENT XV.

THE DENSIMETER.

¶ 39. **Hydrometers and Densimeters.** — There are various kinds of hydrometers employed in the arts.

Nicholson's has been already described, and is the type of a "hydrometer of constant immersion;" that is, one which in use is always made to sink in a liquid to a given mark. A common glass hydrometer is, on the other hand, an example of "variable immersion." The distance it sinks in a fluid depends upon the density of the fluid, and is read by a scale attached to the stem of the instrument. The scales used in the arts are generally arbitrary. The principal ones are those invented by Baumé, Beck, Cartier, and Twaddell, which are compared in Table 40 with a scale of density. The instruments most convenient for scientific purposes carry a scale which indicates at once the density of the liquid, and hence bear the name of densimeters.

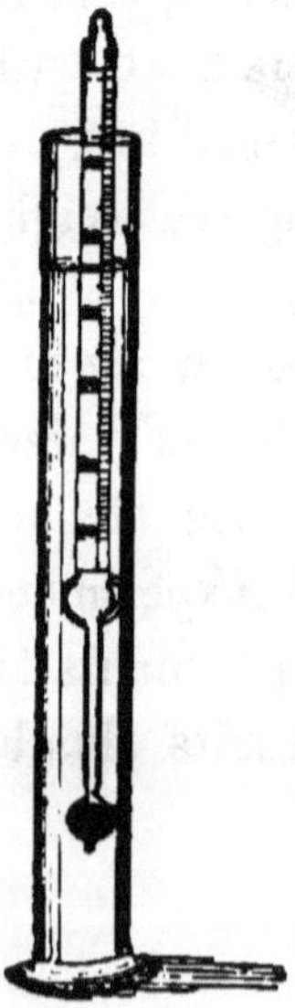

The sensitiveness of a densimeter evidently depends upon the smallness of the graduated stem, compared with the whole displacement of the instrument; but if we make the stem too small, a single hydrometer of the ordinary length can cover only a very limited range of densities. A set of three instruments is often used,—one for liquids lighter than water, one for liquids heavier than water, and one for liquids of intermediate density. There are also sets of twelve or more hydrometers, covering together the whole range of densities from sulphuric acid (1.8) to ether (0.7). With these great accuracy and rapidity may be

Fig. 20.

attained, even without applying any of the ordinary corrections;[1] but if rapidity be the chief object, a single instrument with a "specific gravity scale" will be found most convenient. Such a one is often called by dealers a "Universal hydrometer" (see Fig. 20).

The errors of such instruments are not so great as one might expect, considering that the scales are printed in quantities from originals none too carefully made, fitted to tubes of by no means uniform bore, regardless within certain limits of their size, and fastened to these tubes at a point too high or too low, as the case may be. Still, even if the reading in water is found to be nearly correct, considerable errors may be discovered in other parts of the scale. As these errors depend largely upon the calibre of the tube, the process of correcting them may be properly called calibration (§ 36).

¶ 40. **Calibration and Use of a Densimeter.** — The reading of the instrument is taken while floating successively in at least three standard liquids of known density, such as water, alcohol, and glycerine (see Tables 25–27), then in a number of other liquids whose density is to be determined. As with a Nicholson's hydrometer, the under surface of the liquid is (when possible) used as a sight (see Fig. 6, ¶ 6); and the same precautions are taken to avoid friction against the sides of the jar, and the effects of capil-

---

[1] It should be remembered that changes of atmospheric density influence only that portion of a hydrometer which is above the liquid, and hence will not generally affect even the fourth place of decimals. The effect of a narrow range of temperature in changing the volume of a glass hydrometer is equally unimportant.

lary action due to the stem's becoming dry near the surface of the liquid. Both the densimeter and the thermometer (which is invariably read in every observation) must be washed after immersion in each liquid, either under the faucet or in three changes of water; they should also be carefully dried before immersion in a new liquid; otherwise more or less dilution or mixture is sure to take place. The corrections of the densimeter are then calculated and applied as explained in the next section.

¶ 41. **Treatment of Corrections by the Graphical Method.** — Correction and error are by definition (§ 24) equal and opposite. If the observed value of a quantity is greater than its real value, we say that the error is positive, the correction negative. Thus, by subtracting the observed from the tabulated densities of water, alcohol, and glycerine at a given temperature, we find the several corrections for the instrument by which these densities were observed.

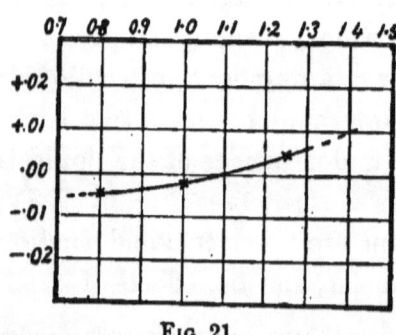

Fig. 21.

The correction of an instrument will generally vary according to the reading in question; hence, to find the correction for every reading, it is necessary to construct either a table of corrections or a curve. Thus, in Fig. 21 the three points indicated by crosses represent (see § 59) corrections of a particular densimeter corresponding to three densities : namely, for alcohol, density 0.80, correction —.004 ; for water, density 1.00, correction — .002 ; for glycerine, density 1.25, correction +.004. The curve drawn by a bent ruler through the crosses enables us to find approximately the correction of this instrument for all intermediate densities by the general rules of the graphical method (§ 59). Thus for an ammoniacal solution of the density 0.9 or thereabouts, the correction would be not far from —.003. Corresponding corrections should be applied to each of the liquids whose density has been determined by means of the densimeter.

EXPERIMENT XVI.

BALANCING COLUMNS.

¶ 42. **Determination of Density by Methods of Balancing Columns.** The ordinary method of balancing columns is illustrated in Figure 22. Some mercury, for instance, is poured into a U-tube, then into the longer arm some water. Suppose the mercury is thus forced up to a level, b, in the shorter arm, and down to a level, c, in the longer arm, by a column of water reaching from a to c, and let the vertical distances, bc and ac, between the corresponding levels be measured ; then since the density of water is known, the density

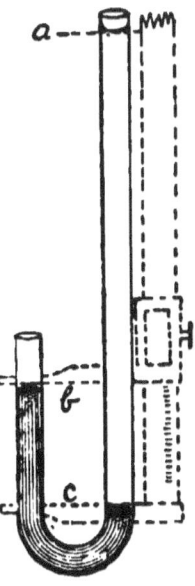

FIG. 22.

of mercury will be determined (see ¶ 43). When the two liquids are miscible this method cannot be applied.

Another method in which this difficulty is avoided is illustrated in Figure 23. A tube in the form of an inverted Y is plunged into two vessels, c, containing water, and d, let us say, glycerine. The two liquids are then sucked up cautiously to the respective levels a and b; and held there by closing a stop-cock in the stem of the Y. The relative density of the glycerine will then be determined by measuring the distances ac and bd. These distances are measured vertically, in the case of each liquid, between its level in the tube and its level in the cistern.

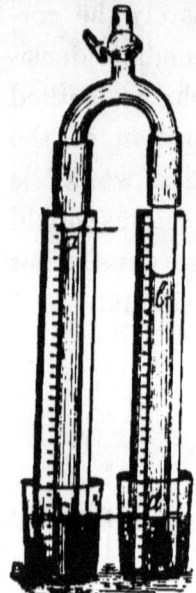

FIG. 23.

For measuring long distances, as ac or bd, a millimetre scale behind the tubes will suffice; for short distance (as bc, Fig. 22) a vernier gauge may be preferable; but special care must be taken to have the shaft vertical. To diminish the effects of capillary attraction, the tubes should have a diameter of a centimetre if possible,[1] and the level should be read

---

[1] If smaller tubes are used, two experiments must be made. In one, the columns of liquid should be as long as may be convenient; in the other as short as possible. The effects of capillary action are then eliminated in the usual manner by taking differences (see § 32). Thus instead of the column ab (Fig. 22) we find the difference between two such columns in the two experiments; and in the same way we find the difference between the two columns bc. These differences evidently balance one another.

by the middle point of the surface, whether convex or concave (see case of the Barometer, ¶ 13).

The common temperature of the two balancing columns may be found by a mercurial thermometer midway between them. An observation of the barodeik will be unnecessary.

¶ 43. **Theory of Balancing Columns.** Two liquid columns are said to balance one another when they exert equal and opposite pressures at a given point. Since pressure is affected by the density as well as by the depth of a fluid, the greater height of one column must counterbalance the greater density of the other. In other words, the densities of two balancing columns must be to each other inversely as their vertical heights.

It is evident that, in Figure 22 of the last article, the vertical height of the water is equal to ac, the total length of the column; but that of the mercury which balances it is only a portion of the whole column of mercury, namely, bc; for the part in the bend of the tube having the same level, c, at both ends, exerts no pressure to the right or to the left (§ 62), and serves simply to transmit pressure from one column to the other. For the same reason, we disregard in Figure 23 the portions of the liquids below c and d, and find that the balancing columns are ac and bd.

The balance between the two liquid columns in the first method (Fig. 22) will not be disturbed by the atmospheric pressure, provided that it affects both columns alike, as is very nearly the case; but strictly we must observe that the barometric pressure is greater at b than at a. There is, as it were, a column of air of the height ab acting on the mercury without any equivalent acting on the water. Since the density of air is about 800 times less than that of water, we should subtract from the apparent length of the column of water, ac, one 800th part of the distance ab, to find the column of water which would balance the mercury *in vacuo*.

In the second method (Fig. 23), supposing c and d to be on the same level, we find in the same way an unbalanced column of air, ab, acting on the shorter of the two columns of liquid. If the longer column is as before, water, we subtract from it one 800th of ab. If the shorter is water we add one 800th of ab. In applying this correction, we neglect the fact that the air within the tube is slightly rarefied, since the accuracy of the instrument employed will not justify more than a rough approximation to the density of the air in question.

If in either method l is the length of the column of liquid whose density, D, is to be determined, w the length of the column of water which balances it *in vacuo*, and d the density of this water at the observed temperature (see Table 25), we have, solving the inverse proportion mentioned above,

$$D = \frac{wd}{l}.$$

EXPERIMENT XVII.

DENSITY OF AIR.

¶ 44. **Determination of the Density of Air.** — A stout flask provided with a stop-cock (Fig. 24) is made thoroughly dry (see ¶ 32), and weighed with the stop-cock open. The flask is then connected with an air-pump, and as much air as possible is exhausted. The stop-cock is now closed; and the flask, having been disconnected from the air-pump, is re-weighed. It should be left on the balance long enough to prove that there is no perceptible gain of weight from leakage of air into it, then quickly opened under water as in Fig. 25. The stop-cock is closed by some mechanical contrivance while the flask is still completely submerged; then the flask is dried outside and weighed with the water which has entered. The temperature of the water is now observed. Finally the flask is filled completely with water and re-weighed. When all these observations have been recorded, an observation of the barodeik (see ¶ 18) is made for purposes of comparison. Having found the proportion of air exhausted, we calculate its density, as explained below.

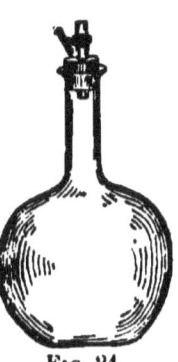

Fig. 24.

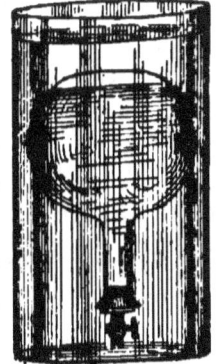

Fig. 25.

¶ **45. Theory of the Partial Vacuum.** — When a flask from which the air has been partially exhausted is opened under water as in Figure 25, the water is forced inwards until the residual air is sufficiently compressed to resist the atmospheric pressure from outside. If the temperature is constant, as will be essentially the case when the flask is surrounded by water, the pressure depends chiefly on the density (see § 78); hence the residual air is compressed until its density is the same as that of the outside air. The space which it then occupies, compared with the whole capacity of the flask, will then represent the proportion of air remaining in it; and the amount of water which enters compared with the total amount necessary to fill the flask will represent the proportion of air exhausted.

The flask must not be plunged too deep below the surface of the water, for if it is the air within it may be perceptibly compressed; but it is well to submerge it to a depth of 10 or 20 *cm.*, to offset the expansion of the air caused by its taking up vapor from the water with which it comes in contact (see Table 13). The less air there is, the less will be its expansion. To obtain accurate results, we must therefore exhaust nearly all the air, or else substitute for water some less volatile fluid.

It may be observed that the water which enters the flask replaces, bulk for bulk, that portion of the air which has been exhausted. The weight of this air is the difference between the weights of the flask before and after exhaustion; the weight of the equivalent bulk of water is the difference between the last two weighings, — before and after the admission of water. We notice that in this experiment, unlike those which precede it, the water enters the flask without displacing any air whatever; hence no allowance is made for the weight of air displaced. Both the weight of air exhausted and that of the water which takes its place are affected by the buoyancy of the atmosphere upon the brass weights (§ 65), and in the same proportion; hence their quotient is unaffected, and represents the true specific gravity of the air referred to the water. This should agree closely [1] with the atmospheric density indicated by the barodeik.

EXPERIMENT XVIII.

DENSITY OF GASES.

¶ 46. **Determination of the Density of a Gas.** — A light flask, as large as the balance pans will admit, is made perfectly dry (see ¶ 32), and weighed with its stopper beside it. To determine the density of the air within the flask, an observation of the barodeik is made (see ¶ 18). Then the flask is filled with coal-gas conducted through a rubber tube reaching as far as possible into the flask. To prevent the escape of the coal-gas, which is lighter than air, the

[1] The true specific gravity of any substance referred to water at any temperature must strictly be multiplied by the density of water at that temperature (see § 69), to find the density of the substance in question. In the present case, the multiplication will hardly affect the last significant figure of the result.

flask is held in an inverted position throughout the process; after which the tube is drawn slowly out of the flask without checking the flow of gas (see

Fig. 26.

b, Fig. 26), and the stopper (a) is immediately inserted. The weight of the flask is again determined. More gas is then passed into the flask as before until it reaches a constant weight. The temperature of the gas in the flask is then found by a thermometer inserted through a bored stopper; and the pressure is determined by an observation of the barometer. Finally the flask is filled with water and weighed for the purpose of finding its capacity.

The last weighing and the observation of temperature which should accompany it may be comparatively rough; but the weighings with air and with gas should be made with the utmost precision, since the difference between them, upon which the result depends, is so slight that even a small error would affect this result in a very considerable proportion (see § 36). If ordinary prescription scales are used, the result should depend upon the mean of at least five double weighings in each case. When great accuracy is desired, a counterpoise should be used consisting of a second flask, hermetically sealed, equal to the first in volume and nearly equal in weight. Small weights added to the counterpoise should bring about an exact adjustment. By using such a counterpoise, changes in atmospheric density are eliminated, since the air will buoy up the contents of both pans alike.

The capacity of the flask is then calculated as in ¶ 32, and the density of coal-gas at the observed temperature and pressure is found by the formula of § 70, using the density of the air indicated by the barodeik. The result is then reduced to 0° and 76 cm. pressure by the formula of § 81.

EXPERIMENT XIX.

MEASUREMENT OF LENGTH.

¶ 47. Selection of a Standard of Length. — A careful comparison of the various scales which we have hitherto employed for the measurement of length will generally show cases of disagreement. These may sometimes be explained as the result of expansion by heat (see Table 8 b); for, though a scale should be correct at 0°, unless otherwise stated, there is no agreement to this effect among manufacturers.[1] In other cases errors are discovered which may be traced to the machine by which the scales are divided. It will not do to assume that the most carefully finished scales are the most accurate. Those printed in large quantities on wood compare very

[1] English measures are generally adjusted (if at all) to a temperature of about 62° Fahrenheit. Certain French manufacturers maintain that all standards are supposed to be correct at 4° Centigrade. In the case of brass metre scales, discrepancies of nearly half a millimetre may sometimes be traced to the temperatures at which they have been adjusted.

favorably with common varieties of "vernier gauge" (see Fig. 27). The latter, in particular, need to be tested as will be explained below. For this purpose, "end standards" are made by various manufacturers with a considerable degree of precision. In place of these, however, the student will find it more instructive to use one depending, as follows, upon his own measurements.

The volume, v, of a glass ball has already been determined (¶ 29); from this the diameter, d, may be calculated by geometry, using the formula [1]

$$d = 1.2407 \sqrt[3]{v}.$$

In calculating the diameter of a sphere from the cube root of its volume, great accuracy may be obtained (see § 36). Thus if the volume is really 40.00 *cu. cm.*, and owing to an error of 1 *cg.* in weighing, the observed value is 40.01 *cu. cm.*, the calculated diameter will be 4.2435 *cm*, instead of 4.2432 *cm*. The difference (.0003 *cm.*) between the calculated and the true value would be imperceptible.

If the ball which we employ is not perfectly spherical, an *average* diameter will be given by the formula. We shall see in ¶ 50, I. how slight irregularities can be allowed for. We may therefore obtain from our experiments in hydrostatics a standard, in the form of a sphere, by which it is possible to correct the reading of a vernier gauge, or any other kind of caliper.

---

[1] This is derived from the ordinary formula —

$$v = \frac{\pi}{6} d^3$$

¶ 48. **Testing Calipers.** A caliper is an instrument intended especially to determine by contact the diameter of bodies, generally the outside diameter. It is provided with two points called "teeth" or "jaws," one of which at least is movable. In one class of calipers the jaws are hinged together, their motion being magnified in some cases by a long index; in another class there is a sliding motion, as in the vernier gauge used in Experiment 1 (see Fig. 27); in a third class the motion is produced by a screw, as in the micrometer gauge (Fig. 28).

The instrumental errors (§ 31) likely to arise differ, of course, according to the special construction of the gauge in question; but there are certain classes of defects common to all calipers, and hence it is well, before beginning any series of measurements, to make a regular examination of each instrument, covering the following points: —

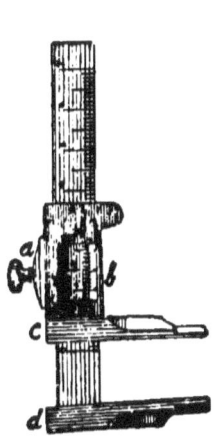

Fig. 27.

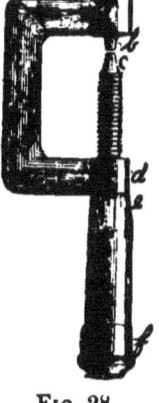

Fig. 28.

(*a*) DISTORTION. The shank of a vernier gauge (*ad*, Fig. 27) should appear perfectly straight to the eye, when "sighted" in the ordinary manner, and perfectly free from twist. A micrometer screw (*cd*, Fig. 28) should similarly appear straight, so that the tooth *c* may be accurately centred in all positions.

(*b*) CONTACT. The jaws of a gauge must be able to touch each other at some point (as *pp'* Fig. 29)

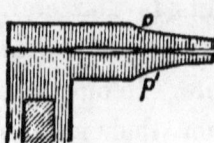

FIG. 29.

convenient for measurement. The shape of these jaws may be modified, if necessary, by the use of a file, or by the application of solder, in order that this condition may be fulfilled. The location of the point of contact is found by examining the streak of light between the jaws.

(*c*) PERPENDICULARITY. The surfaces of the teeth or jaws at the point of contact should be at right angles with the shank of the gauge. In the case of a micrometer, any obliquity immediately appears when the screw is rotated. To detect it in a sliding gauge it is necessary to reverse one of the jaws (as *b* in Fig. 30), and to see whether the two inner surfaces remain parallel.

FIG. 30.

(*d*) GRADUATION. The uniformity of the thread of a micrometer screw is sufficiently established if it turns in the nut, when well oiled, with equal facility throughout its entire length. The graduation of a vernier gauge is most easily tested by the vernier itself; for if the latter always subtends exactly the same number of divisions on the main scale, these may be assumed to be sensibly uniform.

(*e*) LOOSENESS. A gauge should slide freely from one position to another; but any looseness in the moving parts must be prevented. For this purpose a set screw (*a*, Fig. 27) is usually attached to a vernier scale. In the absence of any equivalent arrangement, a nut may often be tightened successfully by pinching it slightly in a vice.

If the defects here mentioned cannot be overcome, the caliper should be discarded for the purposes of the exact measurements which follow.

¶ 49. **Precautions in the Use of Calipers.**

(*a*) WARMTH. In ordinary measurements with a vernier gauge, the warmth of the hand will hardly cause a perceptible expansion; but with micrometers, considerable care must be taken to avoid errors from this source. The usual method is to hold the instrument with a cloth, but it is still more effective to mount it in a vice, and thus to leave both hands free for making the necessary adjustments.

(*b*) CLAMPING. When a caliper has been "set" on a given object, it is customary to clamp it before making a reading, lest in the mean time dislocation should take place. There is danger, however, that in the very act of clamping any instrument, its "setting" may be disturbed. Vernier gauges, unless specially provided with springs to keep the moving parts in place, are troublesome in this respect. The difficulty is lessened by keeping a moderate pressure on the clamp while the setting is taking place. In all instruments, the accuracy of a setting should be tested after clamping.

(*c*) STRAIN. The teeth or jaws of a caliper must obviously not be bent forcibly apart by the pressure between them and the object on which they are set;

for the bending will introduce an error in the reading. One may judge whether the pressure is excessive or not by the muscular force required to produce it, or by the hold which the caliper seems to have upon the object in question. The best micrometers are provided with a friction head (f, Fig. 28) which slips when the required pressure is obtained. A most important result is thus secured, namely, a uniform pressure in all settings of the gauge, including the zero reading (see § 32) whereby the effects of strain may be eliminated.

(*d*) ROUGHNESS. If the surfaces of the teeth or jaws of a caliper are not perfectly smooth and flat,

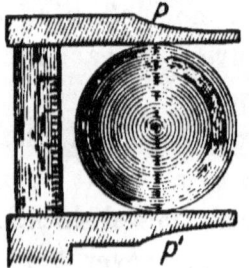

Fig. 31.

an object may fit between them with greater facility in some places than in others. To eliminate the effects of any such irregularity, the diameter which is to be measured should terminate in the points (p and p', Fig. 31) which determine the zero reading of the gauge (see Fig. 29).

These are generally the most prominent points of the inner surfaces; hence the rule, *place the object to be measured where it fits with the greatest difficulty.*

(*e*) OBLIQUITY. The line pp' (Fig. 31) is necessarily parallel to the shank of the gauge; hence also the diameter of any object which coincides with it. If, however, through any mistake in the above adjustment, the diameter to be measured is perceptibly inclined with respect to the line pp', a considerable error is likely to be introduced into the result. It may be shown by trigonometry that if the inclination is less than 1°, the error will be less than one six-thousandth part of the quantity measured; and hence practically insensible. Since the eye can detect under favorable circumstances an obliquity even less than 1°, the following rule will be found sufficiently accurate: *make the diameter to be measured sensibly parallel to the shank of the gauge.*

(*f*) POSITION. An object may be fitted between the teeth of a caliper in various ways, and care must be taken that the diameter thus measured is the one sought. In the case of a rectangular block, for instance, a minimum diameter is usually required, and care must be taken not to place it cornerwise; in the case of a sphere, however, a maximum measurement is wanted, and to secure this, especially when the teeth are rounded (as in Fig. 32), many trials must be made and with the greatest care.

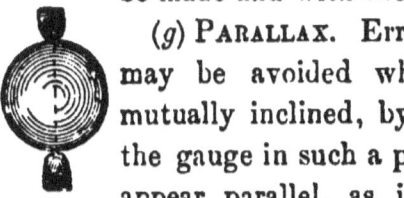

(*g*) PARALLAX. Errors of parallax (§ 25) may be avoided when two scales are mutually inclined, by holding the eye or the gauge in such a position that the lines appear parallel, as in *A*, Fig. 33, not inclined as in *B*.

FIG. 32.

¶ 50. **Correction of Calipers.** —It is important to determine the reading of a gauge or caliper when the jaws are in contact (see Fig. 29). This is called the "zero reading," because it corresponds to a distance zero between the

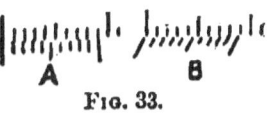

FIG. 33.

points p and p' where contact takes place. A gauge need not be condemned simply because the "zero reading" is not exactly zero. The fulfilment of this condition is in fact exceedingly rare. It is only necessary that the zero reading shall be accurately determined, in order to avoid (by subtraction) all errors from this source (§ 32).

I. VERNIER GAUGE. The general method of reading a vernier gauge has been explained in ¶ 3. We have seen in § 37 how the tenths of the millimetre divisions on the main scale are read by means of a "vernier."

In case, however, the indication of the vernier lies between two numbers, it becomes necessary in all exact measurements to estimate fractions of tenths. We have already found a rough way of representing such fractions (see ¶ 6). A more exact method is described in § 37. To obtain success in applying this method to a vernier reading to tenths of a millimetre, the rulings of the scale should be fine, and a hand lens (such as is represented in Fig. 34) should be used to magnify the vernier and main scale divisions so that the difference between them may be plainly visible to the eye. The student will find it difficult, at first, to select the diagram in § 37 most resembling the case of coincidence in question;[1] but with

FIG. 34.

[1] One of the chief difficulties in conducting this experiment lies in the tendency of students to hold a gauge more or less obliquely, so that all cases of coincidence may appear to be exact, or (what is nearly as hopeless) precisely alike. To an accurate observer, no two settings present in general exactly the same appearance.

THE VERNIER GAUGE.

a little practice most of his errors should be confined to a range of one or two hundredths of a millimetre.

If the zero of the vernier comes opposite a point below the zero of the main scale, the reading is negative. For convenience, however, the negative sign is applied (as in logarithms) only to the whole number indicated on the main scale, — the fraction remaining positive. Thus if the zero on the vernier passes the zero on the main scale by .02 mm. when the jaws are brought into contact, the reading of the vernier should be .98; and in this case, the zero reading is $\bar{1}$.98, according to the general rule given in ¶ 3.

When the zero reading has thus been found within one or two hundredths of a millimetre, a body of

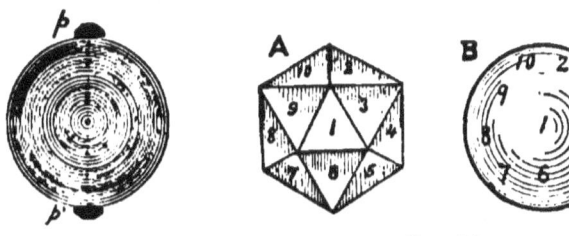

Fig. 35.   Fig. 36.

known diameter is set between the jaws of the gauge. The glass ball, for instance, used in Experiments 8 and 9 is to be placed (see Fig. 31), so as to reach between the points p and p' by which the zero reading was determined (see Fig. 29). Looking at the jaws endwise, we should see the ball symmetrically situated, as in Fig. 35.

If the ball is not perfectly round, we shall need at least 10 measurements of its diameter; and these measurements should obviously be distributed as uniformly as possible over the surface of the sphere. The student will do well to mark in ink ten points upon the ball as in B, Fig. 36, which are to be brought successively under the point p (Fig. 35), in one jaw of the gauge. After each measurement, the corresponding mark should be erased, to prevent confusion. As to the manner of spacing the ten points in question, the student is advised to begin with a 20-sided paper weight (A, Fig. 36), to place a number in the middle of each of the ten faces visible from a given point of view, then to copy these marks on the glass ball B, so that they may appear to be spaced in the same manner in both cases. The geometrician will observe that there is one way and only one way of distributing ten diameters uniformly over the surface of a sphere, and that this way has been here practically adopted.

In each of the ten measurements, a reading is made to hundredths of millimetres; then the zero reading is re-determined. From the mean of the ten measurements above, the mean zero reading is subtracted. We thus find the average diameter of the ball according to the gauge. Dividing this observed diameter by that obtained by the hydrostatic method (which we will suppose to be the true diameter — see ¶ 47), we obtain an important factor, namely, the average space in millimetres occupied by each millimetre division in a certain part of the gauge. If the gauge is uniformly graduated (see ¶ 48, d), it is obviously possible to correct all measurements made with the gauge at the same temperature by means of the factor thus found. In practice, however, it may be assumed that a gauge has been selected in which these corrections are too small to be considered.

II. MICROMETER GAUGE. — In place of the glass ball of Experiments 8 and 9, the student may use the steel balls of Experiments 3 and 4, provided that the displacement of these balls has been confirmed by the specific gravity bottle, as suggested in ¶ 35. The joint volume of these balls is then found by the use of Table 22 (see ¶ 29), then the average volume, from which (the balls being uniform in size) the average diameter is calculated by the formula of ¶ 47.

The diameters of these balls may now be measured by means of a micrometer gauge (see Fig. 28). The tests to be applied to a micrometer and the precautions to be followed are essentially the same as with any other kind of caliper (see ¶ 48 and ¶ 49). The zero reading is found as in the case of a vernier gauge by bringing the teeth into contact. Then the teeth are separated by turning the head of the screw (Fig. 28) until the ball whose diameter is to be measured fits symmetrically between these teeth as in Figure 37.

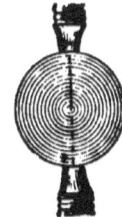

FIG. 37.

The whole number of revolutions of the screw should correspond with the number of main scale divisions on the nut d, uncovered by the barrel e. The hundredths of a turn may be read by the graduation on the edge of the barrel, using as an index a mark running along the nut. Care must be taken to avoid a mistake of a whole turn in reading the gauge; if, for instance, nine whole divisions (nearly) are uncovered by the barrel, and the index points to 98 hundredths, the reading is 8.98 (not 9.98). It is safer with many micrometers to confirm the *whole* number of revolutions by actually counting them.

In reading the micrometer the divisions corresponding to hundredths of a revolution should be divided into tenths by the eye (§ 26). A micrometer with a millimetre thread thus indicates the thousandth part of a millimetre. In the case of a negative zero reading, as with the vernier gauge, the minus sign should be applied only to the whole number of turns.

The diameter of each of the steel balls is determined in this way to thousandths of a turn of the screw; and from the average reading we subtract the average zero reading, observed before and after the above with an equal degree of precision. We find in this way the average number of turns and thousandths of a turn actually made by the screw. Dividing the average diameter of the balls (from the hydrostatic method) by the corresponding number of turns of the screw, we have finally the distance through which the micrometer screw advances in each revolution. This is called the "pitch of the screw." We shall assume that a micrometer has been found, reading to millimetres and thousandths so accurately that in the case of objects of small diameter, no correction need be applied.

EXPERIMENT XX.

TESTING A SPHEROMETER.

¶ 51. **Determination of the Zero Reading of a Spherometer.** A spherometer (Fig. 38) is essentially a micrometer (see ¶ 50, II.) supported by three legs (d, f, g). The vertical screw (ce) has a head (b) divided into a hundred parts, the tenths of which may be estimated by the eye (§ 26). The thousandths of a revolution may thus be read by means of an index (a). This index carries a vertical scale (af), on which the head of the micrometer (b) registers the whole number of revolutions made by the screw. Both on the scale (af) and on the micrometer, the indications should increase as the screw is raised. It is well to renumber the main scale if necessary, so that negative readings may be avoided.

FIG 38.

The zero reading of a spherometer is its reading when the point of the central screw is in the plane of the three feet. To find it, the instrument is set on a piece of plate glass (Fig. 39) of sensibly uniform thickness, selected by the aid of a micrometer gauge, and the screw of the spherometer is raised or lowered until all four points seem to touch the glass at the same time (see Fig. 40). If the central screw is driven too far forward, the instrument will not stand firmly upon the glass, but will have a tendency to rock. This will be noticed especially if one of the feet be held down by the finger, while the other two feet are subjected to an alternating pressure. In fact, the conditions upon which rocking depends are so delicate that a change of a thousandth of a millimetre may cause it to appear or to disappear. When the instrument has been adjusted so that rocking is barely perceptible, the reading is estimated in millimetres to three places of decimals, in the same manner as in the case of a micrometer gauge.

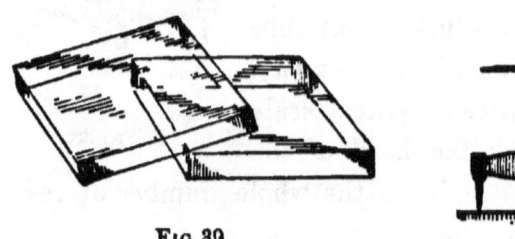

FIG 39.   FIG. 40.

On account of possible irregularities in the glass, at least five readings should be taken in different parts of one surface; and as plate glass is apt to warp slightly in the process of manufacture, five more readings should be taken on the other surface. The mean of the values thus found on a piece of glass of uniform thickness gives the zero reading of the spherometer, and should be determined after as well as before any series of measurements such as will be described in the next section, in order to avoid errors due to change of temperature and to the wearing away of the points upon which the instrument rests.

¶ 52. **Determination of the Pitch of the Screw.** A spherometer with a screw of known pitch can be used in place of a micrometer to measure the diameter of small objects. These are placed upon the plate glass already used to determine the zero reading, and the screw is adjusted so as to touch them from above (see Fig. 41). If the point of the screw is very sharp, and the surface of the object in question convex, great care is needed in finding the maximum diameter.

Fig. 41.

To determine the pitch of the screw, we select an object of known diameter by means of a vernier or micrometer gauge; we may determine, for instance, the diameter of a steel bicycle ball. This is then fitted as above (Fig. 41) beneath the point of the screw, and the reading of the spherometer accurately determined. Subtracting the zero reading, we have the number of turns made by the screw in traversing the diameter of the ball. Dividing this diameter by this number of turns, we have (as in ¶ 50 II.) the pitch of the screw.

Assuming that the screw has a uniform pitch, it is evident that the distance traversed by the point of the screw will always be given by the product of the number of turns and the pitch of the screw.

¶ 53. **Determination of the Span of a Spherometer.** The span of a spherometer, or the average distance of its three feet ($d, f,$ and g, Fig. 38) from the central screw (e) in its zero position (Fig. 40) is an important element in all calculations relating to curvature (see next experiment). It may be determined roughly by a series of measurements with an ordinary vernier gauge. If difficulty is found in measuring directly the distances in question from centre to centre, an impression of the feet and central screw may be taken on paper, and the distances thus indirectly determined.[1] For this purpose the student will doubtless prefer to use a glass scale, if one can be obtained, graduated in millimetres and tenths. In such a scale the rulings should be placed next the paper, and examined with a magnifying glass.

If the feet are blunt (as a and b in Fig. 42), the point of contact will be uncertain. In such a case the feet should be sharpened, and the zero reading re-determined.

Fig. 42.

¶ 54. **Testing a Spherometer.** We have seen that a spherometer may be fitted to a plane surface (¶ 51); in the same way it may be adjusted to a curved surface. To bring this about, the central screw must be driven forward, if the surface is concave, or turned backward if the surface is convex. The distance through which it must be moved obviously depends upon the curvature of the surface in question. The spherometer can therefore be used to determine the curvature of surfaces. There are, however, various sources of error in the use of a spherometer, and to

---

[1] Some authorities prefer not to measure directly the distances (*ed, ef, eg*, Fig. 38) of the three feet from the central screw, but to calculate the span by multiplying the average of the three distances (*df, fg, gd*) between the feet by the square root of one third, or 0.57735.

detect these, the instrument is first of all adjusted to a surface of known curvature, as for instance that of the sphere used in Experiments 8 and 9, (see Fig. 43), or if that is not large enough, to some other sphere of known diameter. The central screw is set as in ¶ 51, so that rocking is barely perceptible, and the reading of the instrument is determined with the same degree of precision as before. At least ten settings should be made on different portions of the spherical surface. In reducing the results we find first the average reading of the spherometer, then subtracting the zero reading we find the number of turns which the screw has made, and hence the distance in millimetres through which the point of the screw has retreated from its zero position, since the pitch of the screw has been already determined in ¶ 52. If this distance is d, and the diameter of the sphere D, the square (s^2) of the span of the spherometer may be calculated by the formula (see ¶ 56, II.), —

FIG. 43.

$$s^2 = Dd - d^2.$$

In this formula, all measurements should be expressed in millimetres. The result should confirm that obtained by squaring the span actually observed in ¶ 53. Slight discrepancies may sometimes be traced to obliquity or excentricity of the central screw, or to irregularities in the shape of the three feet.

EXPERIMENT XXI.

CURVATURE OF SURFACES.

¶ 55. **Determination of the Radius of Curvature of a Spherical Surface.** It is frequently required in optics to know the curvature of the surfaces of a lens; for this curvature, together with the nature of the glass of which a lens is made determines its power of bringing light to a "focus" (§§ 103–104); and conversely, if the curvature and focussing power are known, we may find what sort of glass the lens is composed of. This subject will be fully treated of in Experiments 41 and 42. It is necessary at present only to point out that as the surfaces of lenses are generally ground to resemble portions cut out of a sphere, their curvature may be determined in the same way as that of any other spherical surface.

The spherometer is set upon the lens as in Figure 44, and adjusted so that rocking is barely perceptible as in ¶ 51 and ¶ 54. Ten settings are thus made on each side of the lens, the curvatures of which, even if both are convex, are by no means necessarily the same. Between successive measurements the position of the spherometer should be varied somewhat, so as to determine as well as possible the average curvature of each surface.

FIG. 44.

The results are then averaged for each surface; the mean zero reading subtracted from each, and the distance (d) between the point of the screw and the plane of its three feet thus determined. From this, the diameter, D, of the sphere of which the surface in question forms a part is calculated by the formula (see ¶ 56, I.),

$$D = d + s^2 \div d.$$

where s^2 is the square of the span already calculated in the last article.

The "radius of curvature" is found by halving the diameter.

¶ 56. **Theory of the spherometer.** The formulae of the last two articles depend upon the following considerations: Let a, Figure 45, be the point of the central screw of a spherometer, and b one of the three feet lying in the plane bc, and let ad be a diameter of the sphere abd intersecting the plane bc at c; then if the screw is properly adjusted, acb and bcd will be right triangles. Now abd is

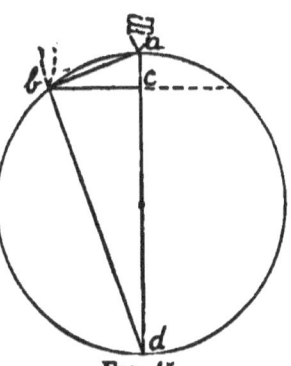

FIG. 45.

also a right triangle, being measured by half the semicircular arc ad; hence the angles cba and bdc are equal, both being complementary to cbd; the right triangles abc and bdc are therefore similar and we have —

$$\overline{ac} : \overline{bc} :: \overline{bc} : \overline{cd}, \text{ whence}$$
$$\overline{cd} = \overline{bc}^2 \div \overline{ac}, \text{ and}$$
$$\overline{ad} = \overline{ac} + \overline{cd} = \overline{ac} + \overline{bc}^2 \div \overline{ac}. \quad \text{I.}$$

We are thus able to calculate the diameter of a sphere (ad) if we know the span of the spherometer

(bc), and the distance, ac, between the point of the screw, a, and the plane of the three feet, bc. We can also calculate the square of the span bc, by the formula, easily derived from the above,

$$\overline{bc}^2 = \overline{ad} \times \overline{ac} - \overline{ac}^2. \qquad \text{II.}$$

EXPERIMENT XXII.

EXPANSION OF SOLIDS.

¶ 57. **Determination of the Coefficient of Linear Expansion.** — By measuring the length of a rod at two different temperatures, the amount of linear expan-

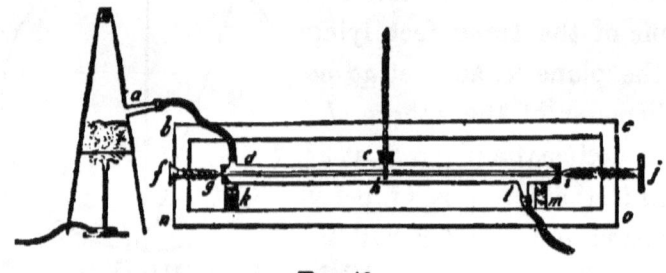

FIG. 46.

sion due to heat may obviously be determined. To make the expansion measurable, a long rod must be employed; and even then delicate instruments are needed to measure the expansion accurately. A micrometer gauge, especially constructed for this purpose, is represented in Fig. 46. It consists of a rectangular wooden frame, $bcon$, capable of admitting a metallic rod, gi, 1 metre long, between the fixed point fg and the point of the micrometer screw, ij.

The rod is surrounded with a tube, also 1 metre long, held in place by the supports, k and m. The tube is closed at both ends with corks, thinner near the middle than at the edges, and serving to keep the rod in position.

A setting of the micrometer is first made with the rod in position, and the reading determined (see ¶ 50, II.); the temperature of the rod is then found by means of a thermometer, h, passing through a cork, e, in the side of the tube. To determine the pitch of the micrometer,[1] it is turned backward (as in ¶ 52) until an object of known diameter fits between it and the end of the rod. A new reading is then made, and the pitch of the screw is calculated as in the case of an ordinary micrometer gauge (¶ 50, II.).

The screw of the micrometer is now withdrawn, to allow room for the expansion of the rod, and steam from a generator (a) is passed through the tube from the inlet (d) to the outlet (l). As soon as a steady current of steam appears at the outlet, a new setting of the micrometer is made.

Subtracting from the last reading of the micrometer the original reading, we find the number of turns made by the screw. From this, knowing the pitch of the screw (¶ 52), we find the expansion of the rod in mm. Subtracting the original temperature (let us say 20°) from the final temperature (100°, nearly, —see, however, Table 14) we find the rise of temper-

[1] By using the same micrometer as in ¶ 52, a determination of pitch will be rendered unnecessary.

ature which has caused this expansion. To find the expansion of 1 *mm.*, we divide the total expansion by the length of the rod in *mm.* (1,000 *mm.*); and we divide the quotient by the rise of temperature in degrees (80° in this instance) to find the expansion in *mm.* of 1 *mm.*[1] for 1°. The result is called the coefficient of linear expansion of the material of which the rod is composed (§ 83).

¶ 58. **Errors in the Determination of Linear Expansion.** — In determining the temperature of a metallic rod by a thermometer beside it, a considerable error is likely to arise unless the temperature of the surrounding air is constant, and the observation prolonged. Air is, as we shall see (Experiment 31), a comparatively poor conductor of heat. To attain greater accuracy in this experiment, the tube may be filled with water, as it is found that an equilibrium of temperature is reached much more quickly with water than with air (see ¶ 65, (6)). A still more accurate method is to replace the tube by a trough packed with melting ice or snow. The mixture should be stirred vigorously for a few minutes, so that the rod may acquire a nearly uniform temperature, not far from 0°. If this method is followed an observation of the thermometer will be unnecessary.

For rough purposes, the temperature of the steam which fills the tube in the second part of the experiment may be assumed to be 100°; but this tempera-

---

[1] The student should note that the expansion of 1 *mm.* in *mm.* is numerically the same as that of 1 *cm.* in *cm.* The result does not therefore need to be reduced to the C. G. S. System.

ture really depends more or less upon the barometric pressure. The thermometer cannot be depended upon to give this temperature correctly, particularly if the bulb only is surrounded by steam. When accuracy is desired, an observation of the barometer must be made (see ¶ 13). The true temperature of the steam may then be found by Table 14, as will be explained in Experiment 25.

It is obviously impossible for the *whole* rod, gi (Fig. 46), to be in contact with the steam or ice surrounding it; for even when the corks are hollowed out, as shown in the figure, so as to leave nearly the whole surface of the rod uncovered, there must still be a small portion at each end which the steam or ice can never reach. The expansion of the rod will not therefore be as great as it should be.

On the other hand, the points fg and ij, being heated by contact with the rod, will expand somewhat, and thus make the expansion of the rod appear to be greater than it really is. To diminish the conduction of heat, the teeth may be protected by the use of insulating material, or by simply pointing them. In all cases contact should be maintained only as long as may be necessary to make a reading of the micrometer. There is always more or less uncertainty as to temperature when a hot and a cold body are in contact. To eliminate errors arising from this source, it would suffice to construct a new apparatus, which should be as short as possible, but otherwise similar to the first, and to calculate the results from the *difference* of expansion in the two cases, according to the general method suggested in § 32.

There is, however, no way to allow for the expansion of the sides of the gauge, caused by the warmth of the steam jacket. We meet here, in fact, one of the fundamental difficulties in the accurate measurement of expansion, — namely, changes in the length of the instruments by which expansion is measured. To avoid errors from this source, a glass tube is sometimes substituted for the metallic tube represented in Fig. 46, so that the expansion of the rod may be observed from a distance. In the most accurate determinations, the gauge or standard used for comparison is insulated from all sources of heat, and even, in some cases, maintained artificially at a uniform temperature.

The expansion of a gauge constructed, like that shown in Fig. 46, principally of wood (see Table 8, *b*), and with sufficient space for the circulation of air, will be found in practice to be very slight; but, in the absence of special precautions, the student should not expect his results to contain more than three significant figures (§ 55).

EXPERIMENT XXIII.

EXPANSION OF LIQUIDS, I.

¶ 59. **Determination of the Coefficient of Expansion of a Liquid by the Method of Balancing Columns.** — A convenient form of apparatus for this experiment

§ 59.] BALANCING COLUMNS. 95

(see Fig. 47) consists of two vertical metallic tubes, ch and fj, about one metre long, with horizontal elbows (cd, ef, hi, and ij) at each end. The lower elbows are connected together with a rubber tube (i), while each of the upper elbows is joined to one end of a differential gauge (ab) by one of the rubber couplings (d and e). Each of the tubes ch and fj is surrounded with a larger tube, or "*jacket*," which can be filled either with melting ice, with water, or with steam. The spouts g and k are to be used either as inlets or as outlets, as the experiment may require.

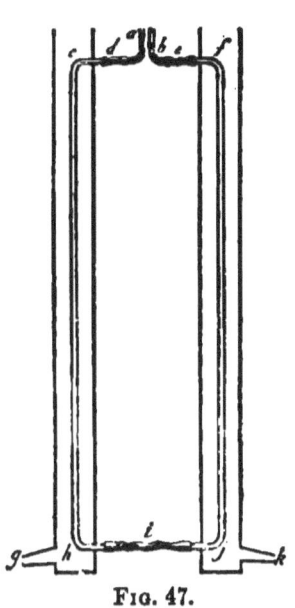

Fig. 47.

The liquid whose expansion is to be investigated is first freed from any air which may be held in solution, by boiling it, then poured steadily through a funnel into the tube a until, after completing the circuit ($adchijfeb$), it issues in a continuous stream from b. The whole apparatus is now inclined first to the right and then to the left, so that any bubbles of air which may be lodged in the horizontal tubes may have an opportunity to escape. A little liquid is next poured out, until the column stands at the level b. This level should be the same, at first, on both sides of the gauge.

Steam is then admitted to the jacket cg through the spout g; and the jacket fk is filled with water from a faucet by a tube connected to the spout k. The temperature of the water is observed after it reaches the top of the tube, f. The height of the liquid in each side of the gauge (a and b) is measured as soon as it becomes stationary, by means of a millimetre scale, as in Experiment 16. (See ¶ 42.) The vertical length of the tube (ch) is finally measured between the elbows (cd and hi), from centre to centre, as close as possible to the jacket. This measurement should (strictly) be made while the tube is still heated by steam.

When the apparatus has become sufficiently cool, the water is emptied out of the jacket fk, which is, in its turn, filled with steam, while the jacket cg is cooled by water from the faucet. The temperature of the water and the reading of the gauge are observed as before; in this case, however, the vertical distance fj is measured. The object of interchanging the jackets is (see § 44) to eliminate errors due to capillarity, or in fact any cause which might tend constantly to raise or lower the level of the liquid on one particular side of the gauge.

Instead of admitting steam to one of the jackets, melting ice may be employed, or water at various temperatures, which must, of course, be observed. The other jacket is always maintained at a temperature not far from that of the room, by the water with which it is filled.

¶ 60. **Precautions in determining Expansion by the Method of Balancing Columns.** — It is evident that the temperatures employed in this experiment must not be higher than the boiling-point nor lower than the freezing-point of the liquid in question, and that this liquid must not be such as to act chemically on the tubes which contain it. Even a very slight action may generate a quantity of gas sufficient to impair the accuracy of the results. The air dissolved in the liquid must be completely boiled out before the experiment, since otherwise bubbles are apt to form when heat is applied. The tubes should be large enough to allow the escape of any air which may be carried into them while they are being filled; but small bubbles can sometimes be dislodged only by jarring the whole apparatus.

The tubes should be completely surrounded with the steam, water, or melting ice by which their temperature is to be regulated. There should be a free vent through one of the spouts (g or k) for the water formed by the melting of the ice, otherwise the temperature of the mixture may rise above 0°. If steam is admitted through one of these spouts, the jacket should be partly covered, leaving only a small opening through which the steam should escape in a slow but continuous stream. If the jackets contain water, the latter should be stirred vigorously to secure a uniformity of temperature. It is well also, in this case, to find the reading of a thermometer at different levels. This will require either a self-registering thermometer, or one with a very long stem. If the temperature is not uniform, the average temperature must be calculated.[1]

The jackets (cg and fk) should be made vertical by a plumb line, as nearly as the eye can judge, and also both branches of the gauge (ab). The tubes cd, ef, and hj should be perfectly horizontal, in those portions at least which are affected by the flow of heat to or from the jackets. The gauge (ab) should be maintained at a uniform temperature (the same always as that of one of the jackets) by surrounding it, if necessary, with water. The tubes of which this gauge is constructed should be of the same uniform calibre, and both perfectly clean, otherwise the effects of capillary action may not be perfectly eliminated. It is well to make sure, both before and after the experiment, that the liquid stands at the same level on both sides of the gauge when the temperature in the two jackets is the same.

To obtain the most accurate readings of such a gauge, a double sight should be employed, as in the case of a standard barometer. The setting is always made so that the plane of the sights may be tangent to the meniscus, or curved surface of the liquid (see ¶ 13 and ¶ 42). The sights may be provided with a vernier reading to tenths of a millimetre.

¶ 61. **Theory of Balancing Columns at Unequal Temperatures.** — The difference in hydrostatic pressure between the two liquid columns, ch and fj, is balanced

---

[1] The average temperature will be indicated at once by an air thermometer of sufficient length, which the student himself may be interested to construct. See Experiment 26.

[61.] COEFFICIENTS OF EXPANSION. 99

by the pressure of a column of liquid reaching from a to b, or more strictly, by the difference between the hydrostatic pressure of such a column and that of an equally long column of air. The latter, being exceedingly light, may be left out of the account. To simplify calculations, we will suppose all the tubes to have a cross-section of 1 $sq.\,cm.$ Then if d is the difference in $cm.$ between the two levels (a and b) in the gauge, when it is maintained at the same temperature (t) as the jacket fj; and if l is the length of the column ch at a higher temperature, t_2; then $l\ cu.\,cm.$ of the liquid at the temperature t_2 plus $d\ cu.\,cm.$ at the temperature t_1, balance $l\ cu.\,cm.$ at the temperature t_1. It follows that $l\ cu.\,cm.$ at $t_2°$ must balance $(l-d)\ cu.\,cm.$ at $t_1°$. Now two columns of liquid of the same cross-section cannot balance one another unless they have the same total weight; hence the same quantity of liquid which occupies $(l-d)\ cu.\,cm.$ at $t_1°$ must expand by the amount $d\ cu.\,cm.$ when heated to $t_2°$, since it then occupies $l\ cu.\,cm.$ If an expansion of $d\ cu.\,cm.$ is caused by a rise of (t_2-t_1) degrees, 1° would cause an expansion in the average (t_2-t_1) times less than $d\ cu.\,cm.$; and since the expansion of 1 $cu.\,cm.$ would be $(l-d)$ times less than that of $(l-d)\ cu.\,cm.$, the expansion (e') of 1 $cu.\,cm.$ for 1° would be

$$e' = \frac{d}{(l-d)\,(t_2-t_1)}. \qquad \text{I.}$$

This expression becomes somewhat modified when the gauge is at the higher temperature, t_2. We have, then, $(l+d)$ *cu. cm.*, all at the temperature t_2, balancing l *cu. cm.* at the temperature t_1. The expansion is as before, d *cu. cm.;* but the quantity expanding is no longer $(l-d)$, but l *cu. cm.* The expansion e'' per *cu. cm.* per degree is therefore

$$e'' = \frac{d}{l\,(t_2 - t_1)}. \qquad \text{II.}$$

We have assumed so far that the tubes have a cross-section of 1 *sq. cm.;* but the principles of hydrostatic pressure are independent of cross-section (see § 63); hence the solutions found in one case may be applied to all. The method of balancing columns is the only one which enables us to measure the expansion of a liquid without taking into account changes in the capacity of the vessel in which the liquid is contained.

The object of this method is to determine an *average* coefficient of expansion between two temperatures rather than the true coefficient of expansion (§ 83) at any particular temperature. The results may differ considerably from those contained in Table 11, which refers in nearly all cases to the expansion of liquids from 0° to 1° Centigrade. We consider, moreover, the expansion of a quantity of liquid measuring 1 *cu. cm.* at the lower of the two temperatures observed instead of at 0°. The result given by the formulæ of this section should, therefore, be designated as *the relative coefficient of expansion from $t_1°$ to $t_2°$*, that is, from the lower to the higher temperature.

EXPERIMENT XXIV.

EXPANSION OF LIQUIDS, II.

¶ 62. **Determination of the Coefficient of Expansion of a Liquid by means of a Specific Gravity Bottle.** — The experiment consists essentially of a repetition of Experiment 14, with a given liquid at two or more different temperatures. These temperatures should be separated from one another as widely as possible, in order that the densities observed may differ by an amount large enough to be accurately measured. The temperatures themselves must be determined with the greatest care, particularly if they are far above or far below the temperature of the room; for in this case rapid changes will take place and must be guarded against.

A convenient way of heating a liquid in a specific gravity bottle to a uniform temperature, is to surround the bottle up to the neck with hot water. To prevent evaporation, the bottle should be closed temporarily by a cork, with a hole made in it sufficiently large to admit freely the stem of a thermometer, to which a brass fan is attached (see Fig. 50, ¶ 65). By this means the liquid is continually stirred until a maximum temperature is reached. As soon as the reading of the thermometer has been observed, the stopper is inserted, with due care not to enclose bubbles of air (see ¶ 32, Fig. 19). The bottle is then carefully dried, and weighed at leisure (see ¶ 33), after cooling to the temperature of the room.

The student is advised not to attempt determinations of density below the temperature of the room, on account of the obvious difficulty of preventing the loss, especially in the case of a volatile liquid, of the portion which is forced out of a specific gravity bottle by its gradual rise of temperature. He should, however, make at least two determinations of density above the temperature of the room, with the liquid already employed in Experiment 14; and he should repeat rapidly the determination made in that experiment at the temperature of the room, to make sure that the result has not been seriously affected by atmospheric changes, or by variations of the density of the liquid due to evaporation or other causes. Coefficients of expansion are then calculated and reduced as explained in the next section.

¶ 63. **Calculation of Coefficients of Expansion.** — Let t_1, t_2, t_3, etc., be the temperatures at which the densities d_1, d_2, d_3, etc., respectively, have been determined and calculated, essentially as in ¶ 38. The results are first represented by points plotted on coordinate paper (see Fig. 48), and connected by a curve drawn with a bent ruler, essentially as in § 59. The necessary forces should be applied to the ruler as near the ends as possible, in order that the curve may be continued downward as far as 0°. The density of the liquid (d_0) at 0° is now *inferred* by means of this curve (see § 59).

The specific volumes, v_0, v_1, v_2, v_3, etc., corresponding to the densities d_0, d_1, d_2, d_3, etc., are now found by the formulæ derived from ¶ 37, —

$v_0 = 1 \div d_0; \; v_1 = 1 \div d_1; \; v_2 = 1 \div d_2; \; v_3 = 1 \div d_3,$

etc. Evidently a certain quantity of liquid expands by the amount $(v_2 - v_1)$ *cu. cm.* when heated from the temperature t_1 to the temperature t_2; that is, $(t_2 - t_1)$ degrees. The expansion per degree is therefore $(v_2 - v_1) \div (t_2 - t_1)$. Since the quantity of liquid

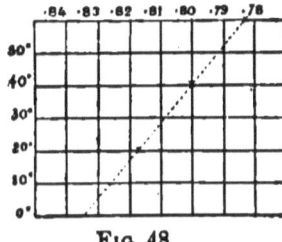
FIG. 48.

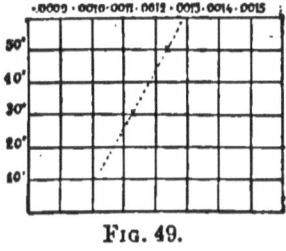
FIG. 49.

thus expanding occupies v_0 *cu. cm.* at 0°, the expansion (e) of a quantity occupying 1 *cu. cm.* at that temperature would be one v_0th as large, or

$$e = \frac{v_2 - v_1}{v_0 (t_2 - t_1)}.$$

The coefficient e which determines the expansion of a quantity of liquid occupying the unit of volume at the standard temperature (0°) is a true as distinguished from a relative coefficient of expansion (see ¶ 61); it expresses, however, the average expansion between the two temperatures t_1 and t_2. We find in the same way the average coefficient of expansion from t_2 to t_3 by substituting, in the formula above, t_2, t_3, v_2, and v_3, for t_1, t_2, v_1, and v_2, respectively. Each result may be represented on co-ordinate paper by a cross, at the right of a point half-way between the two temperatures in question, and under the corresponding coefficient of expansion (see Fig. 49). A line drawn through these points represents approximately the coefficient of expansion at any given temperature. It is clear, however, that with only two determinations of the coefficient of expansion, we cannot tell even whether this line should be straight or curved.

EXPERIMENT XXV.

THE MERCURIAL THERMOMETER.

¶ 64. **Preservation of a Mercurial Thermometer.** — It would seem hardly necessary to point out that a mercurial thermometer is an exceedingly fragile instrument; but in the processes of manipulation about to be described, it is frequently required that a thermometer should be subjected to forces very near the limit of its strength, and which, even in skilled hands, may break it. The student is therefore advised to experiment with thin tubes or strips of window-glass, before attempting the calibration of a thermometer; and to examine the almost microscopic thickness of the glass constituting the bulb, before subjecting it to any considerable pressure. In respect to its resistance to a blow endwise, the bulb of a thermometer may perhaps be compared to the point of a lead-pencil when moderately sharp. In attempting to move the mercury in the thermometer by centrifugal force, the student should limit himself to such velocities as he might give to a palm-leaf fan. More thermometers are broken by suddenly arresting than by suddenly creating the necessary velocity. If a glass thermometer be temporarily mounted on a wooden support, like an ordinary house thermometer, it may be much more roughly treated with the same safety.

The full heat of a flame should never be applied immediately to any glass instrument, since fracture will almost inevitably result. By giving to a flame a waving motion, heat may be applied as slowly as may be desired. As soon as the glass acquires a dull-red heat the danger of fracture is past. There will, however, be no occasion for so high a temperature in the case of a thermometer. The student is particularly cautioned against plunging a cold thermometer into hot mercury,[1] or a hot thermometer into any cold liquid whatsoever.

In applying heat to the bulb of a thermometer, care must be taken not to drive out more mercury than there is room for in the expansion chamber at the top of the instrument. The temperature of the mercury should not be raised above its boiling-point[2] (350° C.) *in any part* of the thermometer; for the pressure of the vapor, being transmitted to the bulb, will be likely to cause an explosion.

¶ 65. **Precautions in the Use of a Mercurial Thermometer.** — (1) TEMPER. — In addition to the dan-

---

[1] The thermometer should be placed in the mercury while cold, and gradually heated with the mercury. On account of its rapid conduction of heat, mercury is more likely to cause fracture than other liquids.

[2] Special thermometers are now constructed so as to read safely as high as the boiling point of sulphur (440° C.).

ger of fracture, the accuracy of a thermometer may be greatly impaired by any wide change of temperature, especially if the change be sudden. After a thermometer is freshly made, there is found to be a gradual contraction of the bulb, which continues perceptibly for months and even for years. This accounts for the fact that nearly all old thermometers stand somewhat too high, although they are not supposed to be graduated until the contraction of the bulb has ceased. The value of a thermometer evidently depends partly on its age or "temper." This value may be completely destroyed by a sudden change of temperature.

(2) CHANGE OF FIXED POINTS. — In fact, when a thermometer is simply heated to the temperature of steam, then cooled as gradually as possible, the readings are almost always affected to the extent of one or two tenths of a degree. In the course of a month the thermometer may return to its former reading, but the change is gradual. It is therefore customary to test a thermometer — in ice, for instance — (see ¶ 69, II.) *after* testing it in steam (see ¶ 69, I.), or in fact after subjecting it to any considerable change of temperature.

(3) CONTINUITY OF THE MERCURIAL COLUMN. — Errors in reading a thermometer frequently arise from a break in the mercurial column, which can be guarded against only by inspection. A slight jarring is usually sufficient to make the column reunite; but when a small bubble of air interrupts the column, or when in the expansion chamber a globule

becomes separated from the rest of the mercury, special precautions are necessary (see ¶¶ 65, 67).

(4) TEMPERATURE OF THE STEM. — To make an accurate determination of temperature with a mercurial thermometer, it is necessary that the mercury, in the stem as well as in the bulb, should be raised to the temperature in question. In a thermometer reading to $-10°$ C., for instance, if the bulb only is heated, the errors, even if the thermometer is correctly graduated, will be as follows: at $50°$, $-0°.5$; at $100°$, $-2°.0$; at $200°$, $-7°.6$; at $300°$, $-17°$; etc. As the temperature rises, more mercury flows into the stem, and it becomes still more important to heat this mercury to the given temperature (see ¶ 84).

(5) UNIFORMITY OF TEMPERATURE. — In nearly all determinations of the temperature of liquids, it is necessary to make use of some stirring apparatus, to secure a uniformity of temperature. A small fan of thin sheet brass is customarily attached to the stem of the thermometer, just above the bulb. The stirring is accomplished by twisting the stem of the thermometer. Special devices are necessary when finely divided substances are employed, though the stem of the thermometer itself may (with due care) occasionally be used, especially in mixtures, as of powdered ice and water, where the resistance will be exceedingly small.

FIG. 50.

(6) TIME REQUIRED. — The length of time required to attain an equilibrium of temperature depends largely upon the conductivity of the surrounding medium, and

upon the degree of accuracy which is aimed at. Let us suppose that a thermometer is taken out of a mixture of ice and water, and placed in air at 32°; if at the end of one minute it rises 16°, that is, half-way towards its final temperature, we may expect it to accomplish in another minute half of what is left, or 8°, according to the general law explained in § 89. The temperatures attained would thus be as follows: in 1 m., 16°; in 2 m., 24°; in 3 m., 28°; in 4 m., 30°; in 5 m., 31°; in 6 m., $31\frac{1}{2}$, etc. At the end of 10 minutes the reading would differ from 32° by only $\frac{1}{32}$ of a degree, a quantity hardly perceptible to the eye on an ordinary thermometer. Now, if the thermometer had been placed in water at 32° instead of in air, the temperature would have reached 16° in a few seconds; and at the end of a minute it would have indicated 32° within a very small fraction of a degree. Again, a mixture of hot lead and cold water may take several minutes before the temperature is practically equalized.

One almost always knows, at least roughly, what the final temperature will be. A useful rule is to observe how long it takes the temperature to reach a point half-way between its original and its final value; then to allow from ten to twenty times as long a time before making a determination of the temperature, according to the degree of accuracy required.

(7) OTHER PRECAUTIONS. — The necessity of shielding a thermometer from radiation has been already alluded to (¶ 15). Delicate thermometers

may be perceptibly affected by mechanical, hydrostatic, or even barometric pressure on the bulb, and by mercurial pressure from within. Such thermometers should be tested both in a vertical and in a horizontal position. Other special precautions will be mentioned as the necessity for them arises.

¶ 66. **Selection of a Mercurial Thermometer.** — For the purpose of calibration, it is best to select a glass thermometer, graduated on its own stem (bc, Fig. 51), in degrees at least 1 $mm.$ long, from 0° to 100° centigrade, with a few divisions above 100° and below 0°. The bulb (ab) should have a volume[1] of nearly 1 $cu. cm.$; and the expansion chamber (c) at the top of the thermometer should have about $\frac{1}{10}$ of this

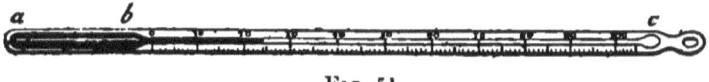

Fig. 51.

capacity. The bulb (ab) should for convenience be elongated as in the figure, so as to pass freely through a hole in a cork fitted to the stem of the thermometer. The expansion chamber should be pear-shaped (see c, Fig. 51), since otherwise particles of mercury are likely to lodge there. The shape and size of the tube must be such that mercury may be made to flow, with a little jarring, from one end to the other; and the quality of the mercury such that there is no tendency for the column to break up into small fragments.

[1] The volume of a thermometer bulb may be estimated by the quantity of water displaced in a small measuring glass (¶ 85). A small bulb usually implies a stem of small calibre, which may give rise to difficulty in calibration.

¶ 67. Manipulation of a Thread of Mercury. — It is frequently required in the calibration of a thermometer to separate from the rest of the mercury in the stem of the thermometer a thread or column of a given length, and to place it in a given part of the stem. When a thread has been broken off, it may be easily moved (by sufficiently inclining or swinging the thermometer) under the influence of its own weight or inertia. For slight motions, jarring is often efficient. The place where the thread breaks off is generally determined by a microscopic bubble of air. To find the location of this bubble, the thermometer is inverted. If a thread of mercury separates at once from the rest, the position of the bubble is evident; if the mercury runs in an unbroken column into the expansion chamber, a small quantity of air will probably be found in the bulb; and if the mercury flows easily back again, there is probably a little air in the expansion chamber.

The (nearly) empty space in the bulb caused by the flow of mercury into the expansion chamber has in any case the appearance of a bubble, which may be made to rise into the neck (b, Fig. 51) by suddenly turning the thermometer into an upright position. If it really contains air, it may be worked up into the stem by jarring the thermometer, especially before all the mercury has had time to flow back from the expansion chamber. If the experiment has been successful, a thread of mercury may now be broken off by inverting the thermometer, and tapping it gently on the table.

In the absence of air in the bulb or in the stem, it remains only to make use of air in the expansion chamber. As much mercury as possible is first made to flow into the expansion chamber, and detached from the rest by jarring the thermometer while in a horizontal position. Then the rest of the mercury is returned to the bulb. If there is any air in the expansion chamber, a part of it will now flow into the bulb; and when the globule of mercury is once more returned to the bulb by centrifugal force (see ¶ 64), a thread of mercury can probably be separated.

The presence of a bubble of air[1] in the neck of the bulb (*b*) greatly facilitates the adjustment of the length of the thread of mercury which will break off when the thermometer is inverted. If the bulb is *slowly* heated or cooled by a certain number of degrees, the mercury will usually flow *by* the bubble without dislodging it, thus lengthening or shortening the thread by that same number of degrees. The surest way, however, of shortening a thread of mercury by a few degrees is to hold the thermometer upright and jar it slightly (see ¶ 64), so that the bubble may rise farther and farther into the stem. If at the same time the bulb is gradually cooled, one may be perfectly sure of shortening the thread to any extent. There is no certain method of increasing the length of a thread of mercury, except by transferring it to the expansion chamber, and adding to

---

[1] Few, if any, thermometers will be found to be entirely free from air.

the globule thus formed more or less mercury from the stem. The globule is then detached and forced backward into the stem, as has been previously described. To prevent it from all returning to the bulb, the latter should be warmed somewhat. The thread will now, probably, be much too long; but may, as we have seen, be shortened at pleasure.

Certain difficulties which are occasionally met in these manipulations may be avoided by the cautious application of heat (¶ 64). It is sometimes impossible to force mercury from the expansion chamber into the stem either through its weight or through its inertia, especially when through accident the expansion chamber has been allowed to become *completely* full. Heat should then be applied to the *top* of the expansion chamber until the mercury is driven out by the pressure of its own vapor. When a thread of mercury can be broken off in no other way, heat may be applied to the stem of the thermometer at the point where a separation is desired. When the mercury refuses to leave the bulb, the flow may be started by slightly warming it; in fact, any desired quantity of mercury may be forced into the expansion chamber in this way (see, however, ¶ 65, (1)).

When the calibration of a thermometer has been finished, as will be explained in the next section, it is well to remove the bubble of air from the mercury. This is done either by cooling the bulb in a freezing mixture (as, for instance, ice and salt) until no mercury remains in the stem; or if this is impossible, by heating the bulb until the air is driven into the expansion chamber. In either case a slight jarring should free the bubble from the mercury. If the bubble is too small to respond to this treatment, it will hardly affect the accuracy of results, unless it actually causes a break in the mercurial column (see ¶ 65, (3)).

¶ 68. **Calibration of a Mercurial Thermometer.** — A thread of mercury, about 50° in length,[1] is placed so as to reach first from 0° upwards, then from 100° downwards. The reading of the end near 50° is taken to a tenth of a degree in both cases, as will be explained below. This enables us to detect any difference in calibre between the upper and lower parts of the thermometer. Next, a thread about 25° long is made to reach first from 0°, then from 50° upwards, then also from 50° and from 100° downwards, with exact readings of the end near 25° or 75°, as the case may be. These will enable us to compare the different quarters of the tube from 0° to 100°. It is not necessary, for most purposes, to carry the process of calibration any further.

To avoid parallax (§ 25) the eye may be held so that the divisions of the scale seem to coincide with their own reflections in the thread of mercury. One end of the thread is always placed so as to coincide exactly with a given division line of the scale (0°, 50°, or 100°), so that any error in the estimation of tenths of degrees will be confined to the reading of the other end. To reduce this error to a minimum,

---

[1] A thread from 49° to 51° will answer. In cases presenting special difficulty, a greater latitude may be allowed.

the student is advised to study or to construct for himself diagrams like the following (Fig. 52), showing the appearance of a mercurial column when dividing the space between two lines into a given number of tenths, and to identify the reading in each case with the diagram which it most resembles.

Before calculating a table of corrections (see ¶ 70) from the results of calibration, it is necessary to determine two "fixed points" on the scale of the thermometer, as will be explained in the next section.

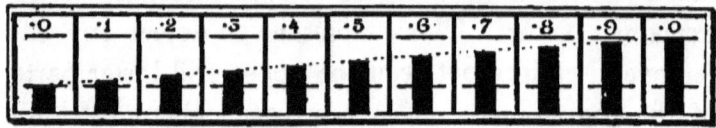

FIG. 52.

¶ 69. **Determination of the Fixed Points of a Thermometer.**[1] — I. The mercurial thermometer is placed in a steam generator (Fig. 53) so that the bulb and nearly the whole of the stem may be surrounded with steam. Only the divisions above 99° project above the cork (*a*) by which the thermometer is held in place. When the greatest accuracy is desired, the sides of the generator are made double, as in Fig. 54. By this means the inner coating, being surrounded on both sides with steam, will have a temperature of 100° nearly, and there will be no radiation of heat between it and the thermometer, since radiation depends upon a difference of temperature (§ 89). It is

[1] The student who is interested in the changes produced in a thermometer by the application of heat will do well to observe the freezing-point before as well as after the boiling-point.

important also to construct a shield of some sort so that the boiling water in the bottom of the apparatus may not be spattered upon the bulb of the thermometer. Such a shield is moreover useful in preventing the thermometer from dipping into the water. It must be borne in mind that the temperature of boiling water is very uncertain, being sometimes

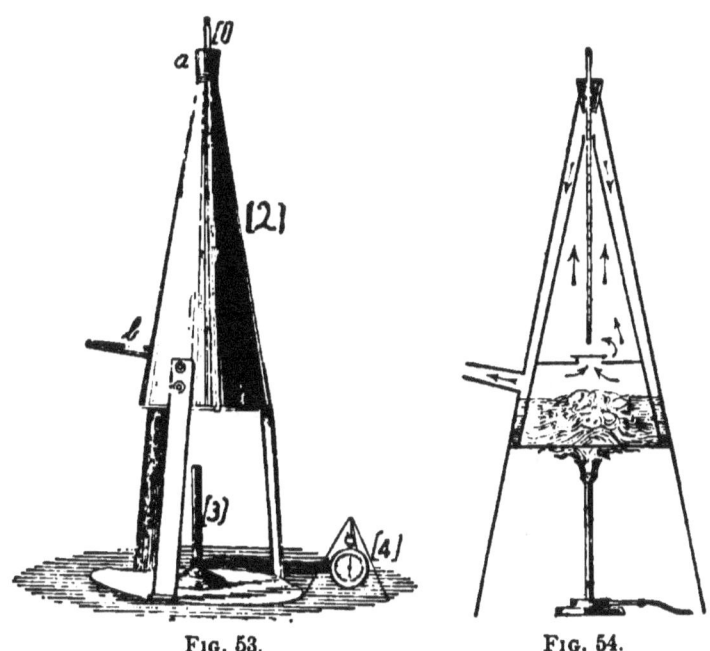

Fig. 53.   Fig. 54.

several degrees above the true boiling temperature, even when the water is perfectly pure, owing to the adhesion of the liquid to the sides of the vessel containing it. On the other hand, the temperature at which steam condenses depends only upon the pressure to which it is subjected.

It is possible, with an apparatus like that shown in Fig. 53, particularly if the spout (*b*) be small, to generate steam so rapidly that the pressure may be perceptibly greater within the generator than it is outside. Care must be taken to check the supply of heat until the feeblest possible current of steam issues continuously from the spout. The atmospheric pressure is then to be observed by means of a barometer ([4] Fig. 53), and the reading of the thermometer determined within a tenth of a degree (see ¶ 68, Fig. 52). If the barometer happens to stand at 76 *cm.*, this reading is called the "boiling-point" of the thermometer, otherwise a correction must be applied, as will be explained in the next section.

II. The thermometer is now allowed to cool as slowly as possible to the temperature of the room, so

Fig. 55.

as not to destroy its "temper" (¶ 65, (1)), then surrounded in a beaker with a mixture of water and finely-powdered ice (Fig. 55), well stirred and covering the scale within one or two divisions of the zero mark. The melting-point of ice is not perceptibly affected by barometric or ordinary mechanical pressure. The ice must be pure and clean. The bulb of the thermometer must not be jammed by the ice (¶ 65, (7)). The reading is to be accurately observed (¶ 68). This reading is called the "freezing-point" of the thermometer.

The boiling and freezing points are called the two "fixed points" of a thermometer, and from them, with the results of calibration, a complete table of corrections should be calculated, as will be explained in the next section.

¶ 70. **Calculation of a Table of Corrections for a Thermometer.** — The correction of a thermometer at 0° is found at once by reversing the sign of the reading in melting ice (see ¶ 69, II., also ¶ 41). If, for instance, the reading in melting ice is $+0°.9$, the correction at 0° is $-0°.9$. The correction at 100° is found by subtracting (algebraically) the actual reading in steam from the true temperature of steam corresponding to the barometric pressure observed. (See Table 14.) Thus if the thermometer reads 99°.0 when the barometer stands at 72 cm., since the true temperature of steam at this pressure is 98°.5, the thermometer stands too high by 0°.5, and the correction is $-0°.5$. It is obvious that under the normal pressure (76 cm.) the thermometer would indicate 100°.5 instead of 100°.0; hence the standard boiling-point is 100°.5 on this thermometer. We find the standard boiling-point in general by adding (numerically) to 100°.0 the correction (at 100°) if the thermometer is found to stand too high, or subtracting the same if the thermometer stands too low.

Let us now suppose that in the calibration of the thermometer a given thread of mercury reached from 0° to 49°.5; if the bottom of this thread had been placed at the observed freezing-point $(+0°.9)$ instead of at the mark 0°, it would evidently have reached farther up the tube. Since the length of the thread can hardly vary by a perceptible amount when it is moved less than one degree, even in a tube with considerable variations of calibre, we may assume that the thread would reach a point just nine tenths of a degree higher than before; in other words, it would reach from 0°.9 to 50°.4. In the same way, if the thread is found to reach from 100° to 50°.7, we infer that it would have reached from the standard boiling-point (found by observation to be at 100°.5) to a point five tenths of a degree above 50°.7, or 51°.2. Between 50°.4, and 51°.2 we find a half-way point[1] on the thermometer, namely 50°.8. If the thread of mercury had been four tenths of a degree longer it would have reached to this half-way point, either from the freezing-point or from the boiling-point. We infer that the volume of the tube included between the boiling and freezing points is exactly halved at 50°.8. Now, by definition, the temperature at which the mercury reaches this point is 50°.0, *according to a perfect mercurial thermometer;* hence the correction for the thermometer at 50° is —0°.8.

In the same way we find the correction of the thermometer at 25°, then at 75°, by considering how far the shorter thread (25° long) would have reached if one end had been placed at +0°.9 instead of 0°, at 50°.8 instead of 50°, or at 100°.5 instead of 100°. We thus find two points near 25°, and half-way between them a third point, showing where the thermometer would stand at a temperature of 25°,

---

[1] This point is sometimes called the "middle point" of a thermometer; but some authorities mean by the "middle point" one *half-way between the divisions numbered* 0° *and* 100° *respectively.*

according to a perfect mercurial thermometer; we find also the indication of the thermometer for a temperature of 75°; and hence also the corrections at 25° and 75°.

The corrections at 5°, 10°, 15°, etc., up to 100° are finally calculated by interpolation. Thus if the correction at 25° is found to be —0°.8, and at 75°, —0°.7, we should find the following table:—

TABLE OF CORRECTIONS.

0°	—0°.9	25°	—0°.8	50°	—0°.8	75°	—0°.7
5°	—0°.9	30°	—0°.8	55°	—0°.8	80°	—0°.7
10°	—0°.9	35°	—0°.8	60°	—0°.8	85°	—0°.6
15°	—0°.8	40°	—0°.8	65°	—0°.7	90°	—0°.6
20°	—0°.8	45°	—0°.8	70°	—0°.7	95°	—0°.5
25°	—0°.8	50°	—0°.8	75°	—0°.7	100°	—0°.5

EXPERIMENT XXVI.

THE AIR THERMOMETER, I.

¶ 71. **Calibration of an Air Thermometer.** — A simple form of air thermometer consists of a glass tube (*ac*, Fig. 56) about 40 *cm.* long, and 2 *mm.* in diameter, closed at one end (*a*). The tube has an

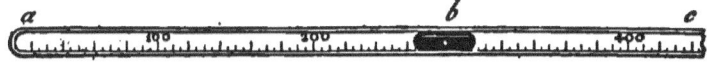

Fig. 56.

engraved millimetre scale, on which an index of mercury (*b*) shows any change in the volume of the enclosed column of air (*ab*). Before closing the end of the tube (*a*), the tube should be thoroughly cleaned and dried.

To test the calibre of the tube, we first weigh it when empty; then we pour in some pure mercury (see ¶ 13) to a depth, let us say, of 5 *cm.*, working it well into the bottom of the tube by means of a fine steel wire. The depth of the mercury is then found as accurately as possible by the millimetre scale, and the tube is re-weighed. Then more mercury is added, a little at a time. After each addition, the depth is recorded, and the corresponding weight is found. This process is continued until the tube is nearly filled with mercury, when the calibration is complete.

Subtracting from each weighing that of the empty tube, we find the amount of mercury contained at each step in the process. Multiplying each weight of mercury in grams by the space in *cu. cm.* occupied by each gram (0.0738 at 20°) we have the capacity of the tube corresponding to the different depths observed. The results are to be entered on co-ordinate paper in the usual method (§ 59). Thus in Fig. 57 the crosses represent volumes from ·1 to ·7 *cu. cm.* corresponding to depths from 0 to 50 *cm.* The curve enables us to find the volume of air enclosed by the index of mercury (*b*, Fig. 56) at any point of the tube. It is easy to show by geometry that unless the crosses all lie in the same straight line, the tube cannot be of uniform calibre.

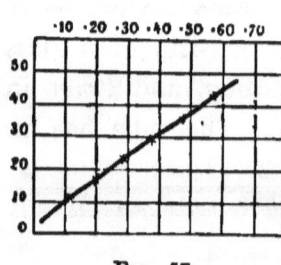

Fig. 57.

¶ 72. Precautions in the Use of an Air Thermometer. —

To obtain accurate results with an air thermometer, it is necessary that the tube should be perfectly clean; for any foreign matter may interfere with the free motion of the mercury index. If in the process of calibration the tube has become coated with the impurities which mercury sometimes contains, it should be scoured with a small wad of cotton on the end of a fine steel wire. Moisture in the tube must be avoided with the utmost care, on account of the vapor which it generates when heated; and in case the slightest trace of condensation appears, the tube must be heated, and dried by a current of air conducted through a still finer tube to the very

FIG. 58.

bottom of the thermometer. The tube must be large enough to allow a free motion to the mercury index, but not so large that bubbles of air may force their way through the mercury.

The mercury used should be of the purest,—at least twice distilled, and perfectly clean and dry. It may be introduced into the tube by means of a medicine dropper drawn out in a flame so as to have a long fine point (Fig. 58). By piercing the mercury, as in Fig. 59, and inclining the tube, the position of the globule may be varied at pleasure. It will be found convenient to place the index so that the lower end may point to a number on the millimetre scale corresponding to the "absolute temperature" (§ 76). Thus if the temperature of the room is 20°, the lower end may be placed at a distance of 273 + 20, or 293 *mm.* from the bottom of the tube. "Absolute temperatures" are indicated approximately[1] by an air thermometer thus constructed; but as the thermometer is affected by barometric changes as well as by changes in temperature, the indications should always be corrected by the method explained in the next section.

To eliminate the effect of the weight of the index, the experiment should be arranged so that the air thermometer may be observed always in the same position. It is necessary, also, that the whole col-

FIG. 59.

umn of air, as far as the index, should be heated or cooled to the temperature which is to be measured. The index must therefore be partly covered in many observations by the heating or cooling apparatus, so that an observation of the upper or outer end will alone be possible. In such cases the length of the index must be allowed for, as what we wish to find is the space occupied, not by the air and the mercury together, but by the air alone. The length of the index must be found by a separate observation in each case, as it is not necessarily the same in different parts of the tube.

[1] Within a few degrees. The air thermometer here described is affected to the extent of about 4° for a rise or fall of 1 *cm.* in the barometer.

¶ **73. Determination of Temperature with an Air Thermometer.** — The reading (r) of an air thermometer is observed, let us say, in a horizontal position, and compared with that of a mercurial thermometer beside it. The air thermometer is then surrounded in a horizontal trough by melting snow or ice, and the reading (r) of the lower end of the index either directly or indirectly determined (see ¶ 72). Then it is surrounded by steam, in an apparatus similar to that shown in Fig. 46, ¶ 57, and the reading (r_1) is again observed. The air thermometer is finally allowed time to cool to the temperature of the room, and again compared with the mercurial thermometer. We will assume, in the absence of any marked change in the barometer or in the temperature of the room, that the air thermometer returns to its original reading, r; if it does not, the experiment should be repeated.

Referring to the curve found in the calibration of the tube (Fig 57, ¶ 71), we now find the volumes v, v_0, v_1, of the confined air corresponding respectively to the observed readings, r, r_0, r_1, of the lower end of the index. The temperature (t) indicated by the air thermometer is then calculated by the formula

$$t = 100 \ \frac{v - v_0}{v_1 - v_0},$$

which is, however, strictly accurate only when the barometer stands at 76 *cm.* (see ¶ 74, VIII.). It is interesting to compare the reading of a mercurial thermometer with the true temperature as indicated by an air thermometer, even if (as will probably be the case) the accuracy of the observations will not justify a correction of the mercurial thermometer.[1] Instead of air, coal-gas or hydrogen may be employed in a thermometer, or in fact any gas not easily liquefied. The results are essentially the same as with the air thermometer. At the same time that air thermometers have for various reasons (see ¶ 74) been adopted as standards of temperature, it is found, by carefully comparing them with mercurial thermometers, that the difference in their indications at ordinary temperatures is generally small in comparison with errors of observation. On account of their greater convenience and precision, mercurial thermometers are therefore employed in most scientific determinations.

¶ 74. **Theory of the Air Thermometer.** — The air thermometer depends upon the Law of Charles (§ 80), that the volume of a gas under a constant pressure is proportional to its "absolute temperature" (§ 76); that is, to its temperature when reckoned from a certain point, about 273° centigrade below freezing, at which it is supposed that all substances would be completely devoid of heat. If T, T_0, and T_1 represent respectively the *absolute* temperature at which the volumes v, v_0, and v_1 were observed, we have, according to the law stated above,

$$T_1 : T_0 :: v_1 : v_0 \qquad \text{I.}$$
$$T : T_0 :: v : v_0 \qquad \text{II.}$$

[1] To lend interest to this experiment, the student may be provided with a very inaccurate mercurial thermometer.

THE AIR THERMOMETER.

From I. and II. we find by one of the ordinary rules of proportion,

$$\frac{T_1 - T_0}{T_0} = \frac{v_1 - v_0}{v_0}, \qquad \text{III.}$$

and

$$\frac{T - T_0}{T_0} = \frac{v - v_0}{v_0}. \qquad \text{IV.}$$

Dividing IV. by III. we have

$$\frac{T - T_0}{T_1 - T_0} = \frac{v - v_0}{v_1 - v_0}. \qquad \text{V.}$$

Now the difference between the freezing and boiling temperatures, T_1 and T_0, under the normal barometric pressure (76 cm.) is divided on the centigrade scale into 100 parts, called degrees, or

$$T_1 - T_0 = 100°, \qquad \text{VI.}$$

and any ordinary temperature, t, is measured by the excess of the corresponding absolute temperature (T) above the freezing point (T_0); that is,

$$T - T_0 = t. \qquad \text{VII.}$$

Substituting the values of $T_1 - T_0$, and $T - T_0$ in VI. and VII. for their equivalents in V., and multiplying by 100°, we have (at 76 cm. pressure),

$$t = 100° \frac{v - v_0}{v_1 - v_0}. \qquad \text{VIII.}$$

If the barometer does not stand at 76 cm. we substitute for 100° in the equation the actual number of degrees between freezing and boiling (see Table 14).

The student may test the accuracy of his work by calculating the "absolute zero" (z), in this case, the temperature at which the index would reach the bottom of the tube, provided that there were no change in the rate at which the air contracts. Substituting in equation VIII. $v = 0$, we have at 76 cm. pressure,

$$z = -100° \frac{v_0}{v_1 - v_0}, \qquad \text{IX.}$$

in which the factor 100° should strictly be corrected as in VIII. for barometric pressure. The meaning of this equation is particularly evident in a special case. If, for example, in a perfectly uniform tube, the index falls from a reading of 373 mm. in steam to a reading of 273 mm. in ice, — that is, 100 mm. for 100°, or 1 mm. per degree, — it is clear that to reach the bottom of the tube it must traverse still farther a distance of 273 mm., corresponding to 273° of the same length. The result of this experiment, when accurately performed with any of the so-called "permanent gases" is invariably to indicate a temperature not far from —273° C. for the absolute zero. It is evident that, if the volume of a gas contracts by an amount equal to one 273d part of its volume at the freezing-point for every degree which it is cooled, the volume will be reduced to nothing at the temperature of 273° below zero; and conversely, if z is the absolute zero, that the gas must gain or lose one zth part of its volume at zero degrees when it is heated or cooled 1° centigrade. The *coefficient of expansion* (e) (§ 83) is therefore numerically equal to $1 \div z$; and may be calculated by the formula

$$e = \frac{v_1 - v_0}{100° \times v_0}. \qquad \text{X.}$$

The coefficient of expansion of all permanent gases is in the neighborhood of .00367.

EXPERIMENT XXVII.

THE AIR THERMOMETER, II.

¶ 75. **Construction of an Absolute Air-Pressure Thermometer.** — A form of air thermometer dependent almost entirely upon pressure is represented in Fig. 60. It consists of a U-tube (abc), with a large bulb (c) blown at the end of the shorter arm, and a somewhat smaller bulb (a) at the end of the longer arm. The apparatus is sealed at the atmospheric pressure with enough mercury to fill the smaller bulb more than half-full.

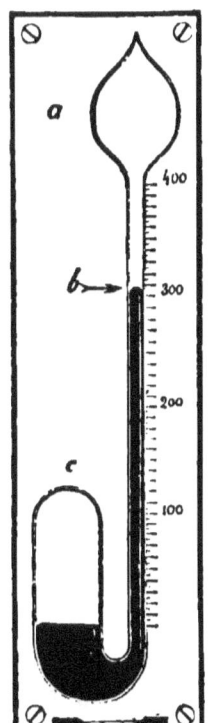

Fig. 60.

It is evident that at the absolute zero of temperature (see § 75), in the absence of any pressure in either bulb, the mercury must stand at the same level in both arms of the U. To locate the absolute zero accordingly, mercury is poured back and forth from one bulb to the other until no difference in the level is observed when the thermometer is returned

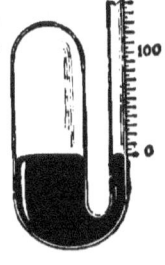

Fig. 61.

to a vertical position. The zero of a millimetre scale is now adjusted to this level (see Fig. 61). By pouring mercury into the bulb *a* (Fig. 60), and suddenly restoring the thermometer to an upright position, the mercury in the tube will be found to stand above its level in the cistern, owing to the compression of air in *c* and its rarefaction in *a*. This process is repeated with more or less mercury in *a* until the column reaches a point *b* on the scale corresponding to the absolute temperature (see ¶ 72). The thermometer should now indicate any temperature correctly on the absolute scale, and has the advantage over that employed in Experiment 26 of being unaffected by atmospheric pressure.

In practice, the bulb *c* is made so much larger than the tube (*b*) that no account need be taken of the variation of the mercury level in *c*. The height of the mercurial column is measured accordingly by a fixed scale. The expansion of the air in the bulb *c* is also disregarded, together with the compression of the air in *a*. All these causes tend to diminish the sensitiveness of the thermometer.

The air thermometer represented in Fig. 60 depends upon the principle (§ 76) that the pressure of a gas which is prevented from expanding increases in proportion to the absolute temperature. When both bulbs (*a* and *c*) contain gas, the pressure in each increases, and hence also the difference in pressure between them increases with the absolute temperature. It follows that the height of the mercurial column which can be maintained by the difference of pressure in question itself varies as the absolute temperature.

¶ 76. **Determination of Temperature by the Pressure of Confined Air.**[1] — A tube (c, Fig. 62), already employed in ¶ 71, is to be connected with a mercury manometer (ab) constructed as follows: two bottles, a and b, are each provided with two siphons passing through an air-tight stopper, one to the top, the other to the bottom of the bottle. The long siphons

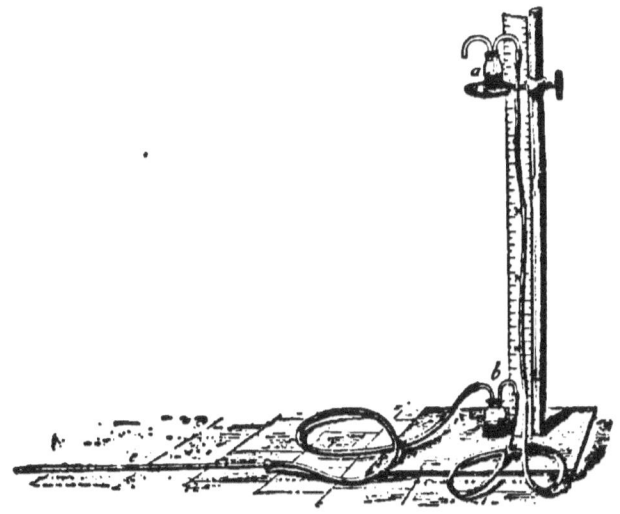

Fig. 62.

and a thick-sided rubber tube connecting them are filled with mercury, and enough more is added to fill both bottles half-full. The mercury stands naturally at the same level in the two bottles; and without disturbing this level, the tube c is connected to the short siphon of one of the bottles, b, by a thick

[1] An experiment illustrating the increase of pressure produced by temperature will be found in Exercise 25 of the "Elementary Physical Experiments," published by Harvard University.

rubber tube, and the reading of the index determined. All the joints must be carefully wound with string to prevent leakage.

The tube c is now surrounded with melting ice, which may be contained in a horizontal trough (see ¶ 57), leaving only the outer end of the mercury index uncovered. The position of the index is then accurately observed. A reading of the barometer is made. The tube (c) is next surrounded with steam, in a steam jacket (Fig. 46, ¶ 57). The air within c is prevented from expanding by raising the bottle, a, on an adjustable platform to a certain height above b (see Fig. 62). The height of b is to be adjusted so that the mercury index in the tube c may stand at exactly the same point as before. The vertical distance between the mercury levels in a and b is then measured with a metre rod. The tube c is now cooled by filling the jacket with water, the temperature of which is to be found approximately by a mercurial thermometer. The height of the bottle, a, is again adjusted so that the index may return to its original position; and the difference between the two mercury levels is measured as before.

Let h_0 be the height of the barometer, h_1 the height of mercury required to prevent the air from expanding when heated to 100° (nearly), and h the height required to confine it at the (true) temperature, t; if we call the pressures of the air v_0, v_1, and v at the absolute temperatures T_0, T_1, and T, respectively; then by definition (§ 74) we have, as in ¶ 74, I. and II.,

$$T_1 : T_0 :: v_1 : v_0 \text{ and } T : T_0 :: v : v_0;$$

¶76.]   THE AIR THERMOMETER.   131

from which we may find, as before, the temperature, t (¶ 74, VIII.), the absolute zero, z (¶ 74, IX.), and a coefficient, e (¶ 74, X.), which determines in this case the proportion in which the *pressure* of confined air increases when heated 1° centigrade. Substituting the values of v_0, v_1, and v, we find

$$t = 100° \frac{h}{h_1} \quad z = -100° \frac{h_0}{h_1} \quad e = \frac{h_1}{100° \, h_0}.$$

It is believed that in the case of a perfect gas the coefficient which determines the increase of pressure per degree should be the same as the coefficient of expansion (Experiment 26). In practice, differences are observed even with the most permanent gases; but these differences are small in comparison with the errors of observation which the student is likely to make.

It is interesting to compare the temperature, t, indicated by an air-pressure thermometer with that indicated by a mercurial thermometer, and to test the accuracy of the work by calculating the temperature (z), at which air would be wholly devoid of pressure, as well as the coefficient e, relating to change of pressure. If the results agree with the values given in ¶ 74, within one or two per cent, the student will be justified in applying a correction to the mercurial thermometer.

EXPERIMENT XXVIII.

PRESSURE OF VAPORS, I.

¶ **77. Application of the Law of Boyle and Mariotte in the Air Manometer.** — One of the most important applications of the Law of Boyle and Mariotte (§ 79) is in the construction of a pressure-gauge, or manometer. A simple form is represented in Fig. 62. It consists of a U-tube, closed at one end and filled with mercury up to a certain level, corresponding to No. 1 on the gauge. The open end of the U-tube is connected with the interior of a vessel, the pressure in which is to be determined. If the mercury stands as before at No. 1, we know that the vessel must be at the ordinary atmospheric pressure. If, however, the air in the closed arm is compressed to half its original volume, we know that the pressure must amount to 2 atmospheres; if the air is reduced to one-third its original volume, the pressure is 3 atmospheres, etc. If, on the other hand, the air expands, the pressure must be less than 1 atmosphere. The pressure in atmospheres may therefore be indicated directly on a scale properly spaced. No. 2 is, for instance, half-way between the closed end of the tube and No. 1; No. 3 is one-third way; No. 4 one-quarter way, etc. Such a gauge is useful in experiments where it is necessary to know roughly the pressure in a closed vessel, as, for instance, a

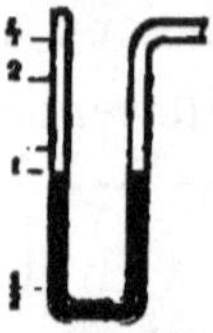

Fig. 63.

¶ 78.] THE AIR MANOMETER. 133

steam boiler. When accuracy is desired, it is necessary to increase the length of the tube, to calibrate it (see ¶ 71), and to allow for the hydrostatic pressure of the liquid in the bend.

The tube already calibrated (¶ 71), for the purpose of measuring the expansion of air, may serve as a manometer. The manometer may be surrounded (if necessary) with water, to prevent the temperature from varying perceptibly in the course of the experiment.

¶ 78. **Testing an Air Manometer.** — The tube (c) is to be connected, as in ¶ 76, with the bottle b (Fig. 62), and the reading of the index determined.

When the bottle a is raised, by means of an adjustable platform, above the bottle b, the air in b, and hence that in c will be subjected to a pressure which can be determined by measuring the distance between the two mercury levels in a and in b by means of a vertical metre rod (see Fig 62). The reading of the manometer c is again determined. The bottle b is now raised above a, so that the air in b and hence also in c will be rarefied by an amount determined in the same way as before. To find the original pressure in c, an observation of the barometer is made (¶ 13).

Let h be the height of the barometer, h_1 that of the column (ab) producing compression, h_2 that producing rarefaction; and let the corresponding volumes of air enclosed by the index in c be respectively (see ¶ 71, Fig. 57) $v, v_1, v_2,$ at the pressures $p, p_1, p_2;$ then evidently $p = h;$ $p_1 = h + h_1;$ $p_2 = h - h_2.$

Now, according to the law of Boyle and Mariotte (§ 79),

$$vp = v_1 p_1 = v_2 p_2;$$

hence we should find

$$v \times h = v_1 \times (h + h_1) = v_2 \times (h - h_2).$$

If these products differ by an amount greater than can be attributed to errors of observation, the determinations upon which they depend should be repeated before making use of the manometer.[1]

¶ 79. **Determination of the Pressure of a Vapor by an Air Manometer.** — The air manometer which has just been tested, is first read at the atmospheric pressure, then connected with a thick rubber tube to

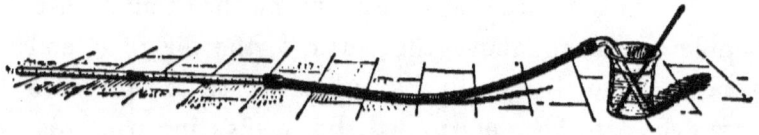

FIG. 64.

a stout tube of glass, closed at one end, and containing ether, already boiling (Fig. 64). The boiling may be effected with safety[2] by hot water, between 50° and 60°. The manometer should be horizontal, but raised somewhat, so that the ether condensing in the rubber tube may run back into the boiler. As soon as the ebullition is checked by the pressure of the vapor generated, an observation of the manometer is made; and at the same time, as nearly as pos-

---

[1] In testing an air manometer from ½ to 2 atmospheres, the errors due to departure from the Law of Boyle and Mariotte will not amount to one fourth of one per cent.

[2] On account of the danger of fire, all flame should be removed from the immediate neighborhood.

sible, the temperature of the water is accurately recorded. When the water has cooled 5°, 10°, etc., new observations of the manometer are made. If the ether ceases to boil, the rubber tube should be cooled, or air let out of it. It is well to put fresh ether in the boiler from time to time. The results are accurate only so long as boiling continues.

The pressure, p_1, corresponding to any reading of the manometer at which the volume, v_1, of air is enclosed, may be calculated from the volume, v, at the atmospheric pressure, p, by the formula expressing the Law of Boyle and Mariotte (§ 79),

$$p_1 = \frac{v}{v_1} p.$$

The results are to be plotted on co-ordinate paper, as explained in § 59, and a curve drawn, as in Fig. 65, to illustrate the pressure of the vapor at various temperatures.

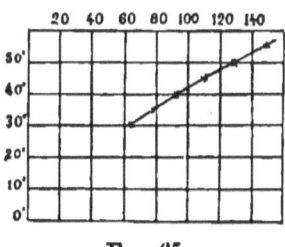

FIG. 65.

EXPERIMENT XXIX.

PRESSURE OF VAPORS, II.

¶ 80. **Dalton's Law.** — We have seen in the last Experiment that the vapor of a liquid may exert a pressure either greater or less than that of the atmosphere, according to the temperature at which the liquid is maintained. The pressure of a volatile liquid is measurable even at the ordinary temperature of the room. To prove this, one has only to inject a few drops of ether with a medicine-dropper, properly bent (see Fig. 66), into the tube of a barometer constructed as in ¶ 13. The ether will form bubbles of vapor even before it rises to the top of the mercurial column; and the pressure of this vapor will cause the barometer to fall some thirty or forty centimetres. By measuring the fall thus produced, the pressure of the vapor of various liquids at different temperatures may be determined.

Another way to illustrate the pressure exerted by the vapor of a liquid is to pour a little of the liquid into a flask, so that it may evaporate into the air which the flask contains.

Fig. 66.

If the flask is corked tightly as soon as the liquid is poured in, a considerable pressure may be generated. In fact, explosions sometimes occur from this cause. To measure the pressure, a tube may be passed through the cork into some mercury in the bottom of the flask (see Fig. 67), and the liquid should be injected by means of a medicine-dropper passing through the cork beside this tube, so as to avoid losing the pressure generated by evaporation before the cork can be put into its place.

It has been found by experiment that the quantity of liquid which evaporates in a flask already containing air, and the pressure which it generates, are exactly the same as in a space from which the air has been completely exhausted. This discovery (known

Fig. 67.

¶ 81.]   DALTON'S LAW.   137

as Dalton's Law) is believed to show that the molecules of a gas occupy very little space in comparison with the space between them, into which a liquid may evaporate. In any case, the height to which the mercury column is raised in Fig. 67 is the same as its depression in Fig. 66, other things being equal. We shall make use of this fact to determine roughly the pressure of a vapor at various temperatures.

We have seen that when a liquid evaporates into a confined space filled with air, the pressure of the air is increased. It is evident that in an open flask the air must expand until the combined pressure of the air and the vapor inside becomes equal to the atmospheric pressure outside. If therefore we know the pressure of the air within the flask, and that of the air outside of it, the difference must be equal to the pressure of the vapor in question. To find the pressure of the air within the flask, it is necessary first to absorb or to condense the vapor which it contains.

¶ 81. **Determination of the Pressure of a Vapor in the Presence of Air.** — To find the pressure of aqueous vapor in an open flask, a small quantity of water is heated in it by submerging the flask up to the neck in a jar of hot water. The temperature of the water within the flask is now determined by means of a thermometer, and a rubber cork is tightly inserted. When the flask has become sufficiently cool it is weighed, then inverted, opened under ice-water, corked, dried, and reweighed with the water which enters it. Finally, it is filled with water and weighed again. A reading of the barometer is made.

Let w_1, w_2, and w_3 be the first, second, and third weights in grams, t the temperature, and h the barometric pressure in cm. within the flask; then the capacity (c) of the flask in $cu.\,cm$. for air or vapor is

$$c = w_3 - w_1 \text{ nearly};$$

and since the volume of air at $0°$ is nearly $w_3 - w_2$ $cu.\,cm.$, its volume (v) at $t°$ is (see § 80)

$$v = \frac{(w_3 - w_2) \times 273 + t}{273}.$$

The pressure of this air at $t°$ is $v \div c$ atmospheres (§ 79), or $hv \div c\,cm$. Hence the pressure (p) of the vapor[1] must be

$$p = h - \frac{hv}{c}.$$

¶ 82. **Evaporation and Boiling.** — The student will notice the regular increase of the quantity of aqueous vapor in the air as the temperature is increased, until finally, as the water approaches its boiling-point, scarcely any air remains in the flask. It is interesting to push the experiment still further, and to expel all the air by actually boiling the water. Boiling may be distinguished from evaporation by the presence of bubbles of pure steam. Unlike the bubbles of air set free from the water by the application of heat, the bubbles of steam may at first completely condense with a crackling sound before reaching the surface of the liquid. When, however, the whole liquid is raised to the boiling-point, the bubbles expand as they escape from the liquid, and if the supply of heat

---

[1] We neglect in this formula the pressure 4.6 mm. of aqueous vapor at 0°.

is sufficient, furnish a steady current of steam which issues from the neck of the flask. The stopper is inserted before boiling has ceased, but, to avoid explosion, not until the source of heat has been removed. When the vapor is condensed by pouring cold water on the bottom of the flask (Fig. 68), ebullition will take place even after the water within the flask is no longer warm to the touch. If the experiment has been successful, a peculiar metallic sound will be heard on shaking the water in the flask. This sound is called the water-hammer, and indicates an almost total absence of air. If the flask is opened under water, it should be completely filled. If opened in air, the space not already occupied by water will be filled with air. The student may be interested to make a rough determination of atmospheric density by weighing the flask before and after the admission of air (see ¶ 44). The capacity of the flask for air is found from the quantity of water which must be added to that already present in order to fill the flask (see ¶ 45). The principal objection to a determination of density by this method lies in the fact that an unknown quantity of aqueous vapor may be taken up by the air which enters the flask. Its advantage consists in the nearly perfect vacuum which is produced by the condensation of aqueous vapor. For further illustrations of evaporation and boiling, see Exercise 22 of the " Elementary Physical Experiments," published by Harvard University.

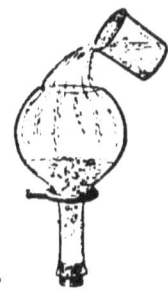

FIG. 68.

EXPERIMENT XXX.

BOILING AND MELTING POINTS.

¶ **83. Determination of Boiling and Melting Points.** — The heater already used to determine the boiling-point of water on a mercurial thermometer may also be employed to find the boiling-points of other liquids. The chief objection to this apparatus is the change of composition which results from boiling away an impure liquid, owing to the fact that the more volatile ingredients are the first to escape. It becomes necessary to condense the vapor before it escapes from the spout, and to make the liquid thus formed return to the boiler. There are, moreover, two practical objections to the use of such an apparatus, — the difficulty of obtaining a sufficient quantity of liquid to fill the boiler, and the danger of fire.

These objections are met by boiling the liquid in a long test-tube, as in Fig. 69. The vapor condenses on the sides but does not escape, and the danger of fire is avoided by the use of hot water instead of a flame as a source of heat.

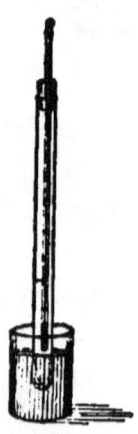

Fig. 69.

Alcohol, for instance, will boil freely if the test-tube is plunged in water at or near the temperature of 100°, since the boiling-point of alcohol is between 78° and 80°. As the water cools it may be used successively to find the boiling-points of chloroform

(58°–61°), bisulphide of carbon (47°–48°), and ether (35°–37°). It is well to have the water about 20° warmer than the boiling-point of the liquid which is to be determined.

The same apparatus, or one with a shorter tube, may be used to determine melting-points. A piece of a paraffine candle may be melted in the test-tube by hot water; then, as it begins to harden, the temperature is observed. Again, by the use of hot water, the paraffine is gradually heated, and the temperature noted at which it begins to melt. Owing to impurity of the paraffine, certain constituents usually congeal more easily than others. It has, therefore, no definite melting point. A certain variety of commercial paraffine melts, for instance, between 53° and 57°. The results are to be corrected as explained below.

¶ 84. **Precautions and Corrections in Determining Boiling and Melting Points.** — To prevent radiation to or from the bulb of the thermometer, and to avoid all danger of spattering (see ¶ 69, I.), a shield may be constructed out of thin sheet brass, small enough to fit into the test-tube. The bulb must not dip into the liquid, but must be surrounded with vapor. The level of the vapor will be distinctly visible through the sides of the tube. It should reach a point a little beyond the end of the mercurial column in the stem of the thermometer, but must in no case reach the open end of the test-tube. A slight escape of the vapor, due to evaporation, cannot be avoided; but a continuous current must be instantly arrested by removing the source of heat.

In finding melting-points, the bulb and stem of the thermometer should be surrounded with liquid up to a point just below the end of the mercurial column. If the stem be dipped any farther into the liquid, it may become impossible to read the thermometer.

The student is advised not to attempt the determination of boiling-points above 100° C.,[1] on account of the danger of accidents. It may, however, be instructive to explain how a temperature above 100° can be determined with a thermometer reading only to 100°. A thread of mercury not over 100° in length is first broken off and stored in the expansion chamber (c, Fig. 51, ¶ 66). The thermometer is then tested in steam (¶ 69, I.). Its reading will be somewhat above 0°; let us say 15°. Then all the readings of this thermometer will be about 85° too low. It is possible, therefore, to determine temperatures up to 185°.

We should, however, remember that a column measuring 85° at a temperature of 100° will measure more or less than that amount, according to the temperature in question. Let the length of the thread of mercury, in degrees, be l, and let the temperature at which this thread is actually observed be t (100° in the instance above); then if t_1 is the temperature to be determined, the correction in degrees is $.00018\,l\,(t-t_1)$. This follows from the value of the coefficient of expansion of mercury; for if a thread

---

[1] Chloroform should be substituted for turpentine (which boils at about 160°) in the second Experiment in Physical Measurement in the list published by Harvard University.

¶ 84.] BOILING AND MELTING POINTS. 143

1° long when heated 1° centigrade expands by the amount 0°.00018, then a thread $l°$ long when heated $(t-t_1)°$ would expand $l \times (t-t_1)$ times as much.

Thus the correction in determining the boiling-point of turpentine (160°) with a thread 85° long, broken off and measured at the temperature 100° instead of 160°, would be $.00018 \times 85 \times (160 - 100)$, or a little over 0°.9. Instead, therefore, of adding 85° to the reading of the thermometer (let us say 74°) we should add, strictly, 85°.9, — that is, the actual length of the thread of mercury at the temperature observed. Instances have already been given (¶ 65, (4)) of errors resulting from heating only the bulb of a thermometer to a given temperature. The corrections in such cases are calculated by the rule given above. That is, we multiply the length of the thread exposed to the air by the difference in temperature between the air and the bulb of the thermometer, to find the correction which should be applied.

In all determinations of temperature, the readings of the thermometer are made to tenths of a degree (¶ 68), and corrected by the table already calculated (¶ 70). The boiling-points of all liquids are affected more or less by atmospheric pressure. A reading of the barometer should always accompany such determinations.

EXPERIMENT XXXI.

METHOD OF COOLING.

¶ 85. Determination of Rates of Cooling. — A calorimeter (Fig. 70) is usually constructed of two (or more) metallic cups, one inside of the other. A vertical section of the calorimeter is shown in Fig. 71, and a horizontal section in Fig. 72. The inner cup, generally made of thin brass, has its outer surface brightly polished to lessen radiation; and for the same reason the outer cup should be polished inside. To prevent the conduction of heat from one

Fig. 70.

Fig. 71.

Fig. 72.

cup to the other, the cups are separated by pieces of cork, which should be sharpened to a point, and held in place by wires. A large flat cork serves to cover both cups, and thus in a great measure to prevent loss of heat; for if the top of the calorimeter were open, a considerable quantity of heat would be carried away by currents of air. In some cases a small stopper is also used, to close the inner cup water-tight.

We prefer for most purposes a calorimeter depending (like that shown above) upon air spaces for its insulation, to one in which these spaces are filled with wool, or other non-conducting material;[1] for though air transmits more heat than wool, it absorbs much less. The heat absorbed by insulating materials is a continual source of error in calorimetry, because there is no simple way of allowing for it. On the other hand, the heat transmitted through the sides of a calorimeter can, as we shall see, be easily determined.

(1) The inner cup is to be filled with hot water, between 90° and 100°, and the temperature of the water is to be found by a thermometer passing through a hole in the cork cover (Fig. 71). The stirrer attached to the stem of the thermometer is used to keep the water in continual agitation; and a stopper is employed to prevent any of it from being spilled over the edges of the cup. Observations of temperature are made at intervals of one minute,[2] and should be continued until the thermometer indicates 30 or 40 degrees. The temperature of the room is then observed; and the quantity of water which has

[1] When no allowance is to be made for loss of heat by the calorimeter, the use of felt is to be recommended. See Experiment 10 in the Descriptive List of Chemical Experiments published by Harvard University.

[2] A clock especially constructed to strike a bell once a minute will be found serviceable in the determination of rates of cooling. Simultaneous observations of time and temperature may thus be made (see § 28).

been used is determined by weighing the calorimeter with and without it.

(2) The experiment is now to be repeated with a much smaller quantity of water, just enough, let us say, to cover the bulb of the thermometer and the stirrer. The calorimeter is to be inclined in every possible direction between the observations of temperature, so as to bring the hot water in contact with every part of the inner cup.

(3) The experiment is again repeated with the same quantity of water as in (2), but without inclining the calorimeter. The stirrer is to be used as in (1), but simply to secure a uniform temperature in the water.

(4) Finally, the calorimeter is to be filled with glycerine or turpentine, warmed by hot water (see ¶ 83). The depth of the liquid, and the method of agitation should be the same as in (1). The temperatures and weights are to be observed as before.

The results of (1), (2), (3), and (4) are to be reduced as will be explained in ¶¶ 86–89.

¶ 86. **Effect of the Temperature and Thermal Capacity of a body on its Rate of Cooling.**
— (1) The results of ¶ 85 (1) are to be represented by a curve (*ab*, Fig. 73), drawn on co-ordinate paper as in § 59. The

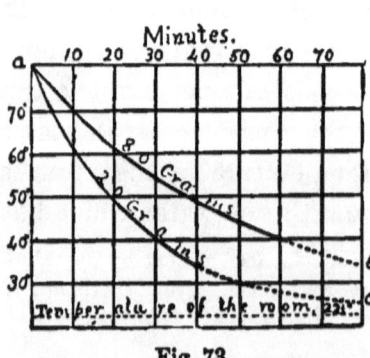

Fig. 73.

scale at the top of the paper corresponds to the number of minutes which have elapsed since the first observation was taken; the scale at the left of the paper represents the observed temperature of the water in degrees. The temperature of the room ($22\frac{1}{2}°$) is shown by the dotted line, which the curve (*ab*) should approach as a limit, — that is, without ever reaching it.

It is advantageous for many purposes that the scale of degrees at the left of the paper should represent, not the temperatures actually observed, but the differences between those temperatures and that of the room;[1] since the rate of cooling depends upon the differences in question (see § 89). If this method is adopted, the first observation should be one about 50° above the temperature of the room.

In any case the student should satisfy himself that Newton's Law of Cooling (§ 89) is approximately fulfilled.[2] Thus the calorimeter may cool (see *ab*, Fig. 73) between the 5th and the 10th minute from 75° to 70°, that is, 5° in 5 minutes; while between the 50th and the 60th minute it may cool only from 44° to 40o, or 4° in 10 minutes. In the first case, when the average temperature ($72\frac{1}{2}°$) is 50° above that of the room ($22\frac{1}{2}°$) we have a rate of cooling equal to 1° per minute; in the second case, with an average temperature (42°) nearly 20° above that

---

[1] This method of plotting the curves must be adopted if the temperature of the room varies considerably in the course of the experiment (¶ 85).

[2] Departures of 20% have been observed in a range of 60°. See Everett's Units and Physical Constants, Art. 143.

of the room, the rate of cooling is $\frac{2}{5}$° per minute. Obviously,

$$50 : 20 :: 1 : \tfrac{2}{5}.$$

In the same way, with 20 grams of water in the calorimeter, the rate of cooling should be found to vary in proportion to the excess of temperature above that of the room. The rate of cooling is, however, very different in different cases, as it depends upon the quantity of water which the calorimeter contains. Let us next consider the relation between this quantity of water and the rate of cooling.

(2) The fundamental principle underlying all determinations by the method of cooling is that the *number of units of heat* (§ 16) lost by a calorimeter per unit of time is proportional to the *difference in temperature* between the inner and outer cups. It does not, therefore, depend upon the contents of the calorimeter except in so far as the nature or quantity of these contents may modify the temperature of the inner cup.

Let us first suppose that in both experiments, ¶ 85 (1) and (2), the water is agitated sufficiently to bring it in contact with every portion of the inner cup, so that a perfectly uniform temperature is the result; then if the outer cup is unchanged in temperature the *flow of heat* from one cup to the other corresponding to a given reading of the thermometer must be in both cases the same. How, then, do we account for the marked differences which we observe in the *rates of cooling?*

The supply of heat in a calorimeter may be compared to the quantity of water in a leaky pail. Given the rate of the stream flowing out of the pail, the time it takes for the water-level to fall one inch is evidently proportional to the horizontal section of the pail. In the same way, with a given flow of heat from a calorimeter, the time required for the temperature to fall 1° must be proportional to what we call the *thermal capacity* (§ 85) of the calorimeter and its contents.

It is obvious from Figure 73 that with 80 grams of water the cup must cool more slowly than with 20 grams. In the first case it takes, for instance (see ab, Fig. 73), 60 minutes to cool from 80° to 40°; if in the second case only 20 minutes are required to cover the same range of temperature, the natural inference is that the thermal capacity in the first case is to that in the second case as 60 is to 20, or as 3 is to 1.

The thermal capacity in question is in no case simply proportional to the quantity of water which the calorimeter contains; for the inner cup, the thermometer, and the stirrer all possess a certain capacity for heat. We may estimate this capacity roughly by the method of cooling. Let us call it c. Then in the first case the total thermal capacity is $80 + c$; and in the second case it is $20 + c$; hence we have

$$80 + c : 20 + c :: 3 : 1,$$

a proportion which can be satisfied only if $c = 10$. We infer, therefore, that the calorimeter, thermometer, and stirrer are together equivalent, in thermal capacity, to about 10 grams of water.

We may assume provisionally that this inference is correct; but for accurate calculations, we prefer a determination of thermal capacity made as will be described in Experiment 32.

¶ 87. **Calculations concerning Loss of Heat by Cooling.** — We have found in the last section (¶ 86, 1), that when a certain calorimeter contains 80 grams of water at an average temperature 50° above that of the room, the rate of cooling is 1° per minute. We have also found (¶ 86, 2) that the calorimeter itself is equivalent in thermal capacity to about 10 grams of water; hence the total thermal capacity is $80 + 10$, or 90 units. The heat lost under these conditions is therefore 90×1, or 90 units per minute. Let us now suppose that the average temperature is only 1° above that of the room, instead of 50°; then by Newton's Law (§ 89) the rate of cooling will be $\frac{1}{50}$ of 1° per minute; hence the loss of heat will be $90 \times \frac{1}{50}$, or 1·8 units per minute.

It follows from the fundamental principle of the method of cooling (¶ 86, 2) that the loss of heat at a given temperature is the same, no matter what substance or substances the calorimeter may contain, provided that every part of the inner cup is brought in contact with the mixture. The rate of flow corresponding to difference in temperature of one degree between the inner and outer cups is accordingly an important factor in calculations (see ¶ 93, 3) relating to loss of heat by cooling.

Unless the calorimeter is filled, as in ¶ 85 (1), or its contents sufficiently agitated, as in (2), the inner cup will not be uniformly heated throughout. When a glass vessel is used (as in Exp. 38), only those portions nearest the liquid may be perceptibly warmed or cooled by it; and even with metallic vessels, especially when thin, differences of temperature can frequently be recognized by the touch. The result is a considerable diminution in the rate of cooling. To estimate the effect in question, we may utilize the results of ¶ 85 (3).

From these results the curve ac (Fig. 73) is to be plotted in the same manner as ab (¶ 86, 1). If in both curves (as in Fig. 73) the first observation utilized is about 80°, we shall find a point of intersection, a, nearly opposite 80° and 0 minutes. We may notice that ab takes 60 minutes to fall from 80° to 40°, while with ac only 30 minutes are required; hence the rate of cooling represented by ac is twice as great as in the case of ab, so that when reduced to 1° difference in temperature, it will be $\frac{2}{50}$ of 1° per minute. Now let the weight of water be 20 grams; then since the calorimeter is equivalent to 10 grams,[1] we have a total thermal capacity of 30 units. The loss of heat is therefore, not 1·8, as before, but $30 \times \frac{2}{50}$ or 1·2 units per minute.

These figures are sufficient to show the importance, in the method of cooling, of comparing two quantities

---

[1] We should remember, strictly, that if only a portion of the inner cup is heated, the thermal capacity will be somewhat less than 10 units.

under exactly the same conditions. Let us suppose that we were to calculate the thermal capacity of the calorimeter from the results of ¶ 85 (1) and (3), in which the conditions are not the same. Since the rate of cooling is twice as great in (3) as in (1), we might infer that the thermal capacity of the calorimeter with 80 grams was twice that with 20 grams. This would make the thermal capacity of the calorimeter alone 40 units instead of 10 (see ¶ 86, 2).

¶ 88. **Construction of a Series of Temperature Curves.** — From an extended series of results[1] it would be possible to construct a series of curves similar to those shown in Fig. 74. It is not, however, necessary that each of these curves should be the result of observation. From two of them, the rest may be obtained with more or less accuracy by different processes of interpolation.

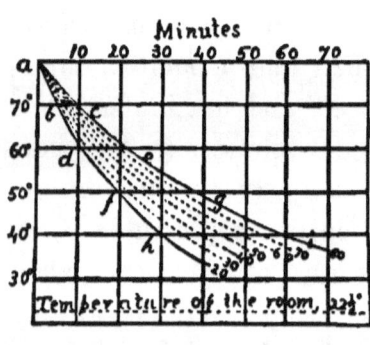

Fig. 74.

Let $acegi$ and $abdfh$ be the two curves already obtained (see Fig. 74), corresponding respectively to 80 grams and to 20 grams of water, and let it be required to draw a curve corresponding to 50 grams of water. Then since 50 is midway between 80 and

---

[1] The teacher may, for the sake of illustration, have a series of curves constructed from the results of a large class of students using different quantities of water.

20, the curve in question may be placed (roughly) midway between the other two; and in the same way other curves may be drawn so as to divide the distance equally into still smaller parts. This method of interpolation is, however, obviously inaccurate, and especially so between such wide limits.

A more accurate method depends upon the principle (see ¶ 86, 2) that the time of cooling is (other things being equal) proportional to the thermal capacity of the calorimeter and its contents. Since 80 grams require, for instance, 10 minutes to cool from 80° to 70°, and 20 grams take only five minutes (see Fig. 74), we may infer that 50 grams would require $7\frac{1}{2}$ minutes; or in other words, that the distance bc would be bisected by the 50-gram curve. In the same way the other horizontal distances, de, fg, hi, etc., would be bisected. To obtain the intermediate curves, accordingly, the *horizontal* distances, bc, de, fg, etc., are each to be divided into a given number of equal parts. The curves may then be drawn through the points of division.

It is easy to show that this method of interpolation, though more accurate than the first, may still lead to considerable errors, when we consider differences in the flow of heat from the calorimeter. With 80 grams of water, 1° above the temperature of the room, we have calculated that the loss of heat amounts to 1·8 units per minute (see ¶ 87); with 20 grams we have found similarly 1·2 units per minute. Let us assume that with 50 grams the loss is midway between these two numbers, or 1·5 units per minute. Then since the total thermal capacity is 60 units, the temperature must fall at the rate of $1{\cdot}5 \div 60$ or $\frac{1}{40}$ of $1°$ per minute. The time required to fall $1°$ at this rate would be 40 minutes; in the case of 80 grams it would be 50 minutes (see ¶ 87); in the case of 20 grams it would be 25 minutes. The times required for 80, 50, and 20 grams to fall through a given range of temperature would be, accordingly, proportional to the numbers 50, 40, and 25, respectively. Since 40 is by no means midway between 50 and 25, the 50-gram curve must be considered as only approximately bisecting the horizontal distance between the other two.

It is evident that if the system of curves shown in Fig. 74 were to be relied upon for exact calculations, it would be necessary to confirm the position of the 50-gram curve, at least, by direct observations. As a matter of fact we shall refer to Fig. 74 only for the purpose of making small corrections for cooling; so that we may disregard any errors in these curves which are likely to arise from an interpolation depending upon a division of horizontal distances into equal parts.

¶ 89. **Calculation of Specific Heat by the Method of Cooling.** — I. A set of curves is to be constructed essentially as in ¶ 88, using, however, in connection with the curve *acegi* (Fig. 74) representing the results of ¶ 85 (1), a curve *abdfh*, derived from the results of ¶ 85 (2), and not (as in Fig. 74) from the results of ¶ 85 (3). The intermediate curves will then represent rates of cooling corresponding to different quantities of water when brought in contact with every part of the inner cup. The results of ¶ 85 (4) are next to be plotted *on tracing-paper*, with a horizontal line (as in Fig. 73 to) represent the temperature of the room. This line is then superposed (by moving the tracing-paper) over a similar line in the new series of curves; and at the same time the curve on the paper is made to pass through the common point of intersection of the series in question (see *a*, Fig. 74).

A curve thus obtained with, let us say, 75 grams of turpentine, may be made to coincide, not with the 70-gram curve, nor with the 80-gram curve (see Fig. 74), but with one rather which would correspond to 30 or 40 grams of water. Under the conditions of the experiment, the heat lost by the calorimeter must be the same whether it contain turpentine or water (see ¶ 86, 2); hence equal rates of cooling imply equal thermal capacities (*ibid.*). Since the calorimeter has the same total thermal capacity with the turpentine as with the water, the 75 grams of turpentine must be equivalent to 30 or 40 grams of water; and 1 gram of turpentine must be equivalent to a quantity of water between $\frac{30}{75}$ and $\frac{40}{75}$ of a gram ; or let us say 0.4 + grams. In other words, the specific heat (§ 16) of turpentine must be 0.4+. In the same way the specific heat of any other liquid might be calculated.

It is evident that the curves of ¶ 88, if thus treated, would not have given an accurate result. 20 grams of water might be found, for instance, under the conditions of ¶ 85 (3), to cool as slowly as the 75 grams of turpentine in ¶ 85 (4); but this would be due, not simply to the fact that water has a greater thermal capacity than turpentine, weight for weight, but also to the fact that a much smaller amount of surface is heated by the water. Obviously the 20 grams of water cannot be equivalent in thermal capacity to the 75 grams of turpentine, because their rates of cooling, though equal, have been compared under dissimilar conditions.

II. Another method of calculating specific heat depends upon a comparison of the rates of cooling of two liquids when *equal volumes* are employed. Let us suppose that the time occupied by 75 grams of turpentine in cooling from 80° to 60° in ¶ 85 (4) is really the same as that of 20 grams of water in ¶ 85 (3),— that is, 10 minutes (see *ac*, Fig. 73),— while that required in ¶ 85 (1) for 80 grams of water (see *ab*, Fig. 73) is 20 minutes; then since the conditions are nearly the same in (1) and (4), the total thermal capacities in question must be to each other as 10 is to 20 (¶ 86, 2). If the calorimeter is equivalent (see ¶ 86, 2) to 10 grams of water, we have with 80 grams of water a total thermal capacity of 90 units; hence with the turpentine the total thermal capacity must be $\frac{10}{20}$ of 90 units, or 45 units. Subtracting from the 45 units the 10 units due to the calorimeter, we find a remainder of 35 units, which must be the thermal capacity of 75 grams of turpentine. Hence the specific heat of turpentine is $35 \div 75$, or $0.4 +$.

The method of cooling has been applied to the determination of the specific heats of solids in the form of powder, as well as to liquids; but it is generally thought to be less reliable than the methods of mixture about to be described (Exps. 33 and 34).

EXPERIMENT XXXII.

THERMAL CAPACITY.

¶ 90. **Determination of the Thermal Capacity of a Calorimeter.** — (1) We have already seen that the thermal capacity of a calorimeter may be calculated roughly from data obtained by the method of cooling (see ¶ 86, 2); but that a very slight change in the conditions of the experiment may make the result worthless. For this reason the method of cooling is hardly to be counted as a practical method for finding the thermal capacity of a calorimeter. The experimental determination of thermal capacity may be made by either of the following methods: —

I. The whole calorimeter is to be weighed, including (see ¶ 85, Fig. 71) the inner and outer cups, the cork supports and cover, and the thermometer and stirrer. The temperature of the inner cup is now found by observing the thermometer, after it has remained within this cup for some time (see ¶ 65, 6). Then water at an observed temperature, between 30° and 40°, is poured rapidly (¶ 92, 4) into the cup until it is nearly full (¶ 92, 8). The cork is immediately inserted (¶ 92, 6) and the time noted (¶ 92, 9). The water is then stirred (Fig. 50, ¶ 65) by twisting the stem of the thermometer, until two successive observations of the thermometer a minute apart (see ¶ 92, 10) agree as closely as in ¶ 85 (1), at the same temperature (see ¶ 92, 8). The resulting temperature is then observed, and the time again noted (¶ 92, 9). The whole apparatus is then re-weighed to find how much water is in the calorimeter (see also ¶ 92, 5).

There are two practical objections to the method just described: first, that the change in temperature of the water is almost too small to be measured accurately with an ordinary thermometer; and second, that the quantity of heat absorbed by the calorimeter may be small in comparison with that lost by cooling (¶ 93), which can only be roughly allowed for.

The change of temperature of the water may be increased by using a smaller quantity of it; but this is objectionable, as will be seen by comparing the results of ¶ 85, (2) and (3), unless the water can be *well shaken* in the calorimeter, or unless the object of the experiment be a determination of thermal capacity of the calorimeter when *partly full*. A thermometer graduated to tenths of degrees will be found useful in this and other experiments where it is necessary to measure small changes of temperature.

II. Another method of finding the thermal capacity of a calorimeter consists in heating the inner cup instead of the water. This may be done by filling the cup with hot lead (or better, copper) shot, the temperature of which is to be determined by two or three observations of a thermometer at intervals of a minute (see ¶ 92, 10). The shot must be well shaken between these observations, to secure a uniformity of temperature (see ¶ 92, 8); it is then poured out, and immediately replaced by water at an observed temperature near that of the room. The resulting temperature is then determined, and the weight of water used is found as before.

The change in temperature of the water may be made practically five or ten times as great in II. as in I., and the correction for its cooling will be comparatively slight. The principal source of error in this experiment is the rapid cooling of the inner cup while empty (see ¶ 92, 4).

(2) The results of an experimental determination of thermal capacity should in all cases be confirmed by a calculation based upon observations of the weights and specific heats of the substances employed in the construction of the calorimeter. The inner cup is to be weighed, also the stirrer (Fig. 50, ¶ 65); and the amount of water displaced by the thermometer is to be found by the aid of a small measuring-glass (Fig. 75). The glass should be filled with water so that the thermometer may be immersed to the same depth as when it is used to determine the temperature of liquids in the calorimeter. The level of the water is then carefully observed with and without the thermometer. It will be assumed that the thermometer is constructed of glass and mercury;

FIG. 75.

the calorimeter and stirrer of brass; otherwise the materials in question must be noted. From these data the thermal capacity of the calorimeter may be calculated (see ¶ 91, III.).

¶ 91. **Calculation of Thermal Capacity.** — We have already considered a method by which thermal capacity may be roughly computed through a comparison of rates of cooling (¶ 86, 2). This section relates to the calculation of thermal capacity from the observations made in ¶ 90.

If, as in the first method (¶ 90, I.), t_1 is the original temperature within the calorimeter, w the weight of water used, t_2 its temperature just before it is poured into the calorimeter, and t the resulting temperature, then, since w grams of water cool $(t_2 - t)$ degrees by coming in contact with the calorimeter, they must give up to it $w \times (t_2 - t)$ gram-degrees, or units of heat (§ 16). This raises the temperature of the calorimeter $(t - t_1)$ degrees; hence to raise it 1° would require a quantity of heat, c, given by the formula

$$c = \frac{w \times (t_2 - t)}{t - t_1}.\qquad\text{I.}$$

This is, by definition (§ 85), the thermal capacity of the calorimeter. To find the temperatures t and t_2, at the time when the water is introduced into the calorimeter, allowances for cooling should be made (see ¶ 93).

The second method (¶ 90, II.) differs from the first in that w grams of water are *warmed* $(t - t_2)$ degrees, and hence must *receive* $w \times (t - t_2)$ units of heat from the calorimeter, the temperature of which is thereby *reduced* $(t_1 - t)$ degrees; hence to reduce it 1° would require a quantity of heat, c, given by the formula

$$c = \frac{w \times (t - t_2)}{(t_1 - t)}. \qquad \text{II.}$$

This formula is evidently reducible to the same form as I.

In the last method (¶ 90, 2) if w_1 is the weight of the inner cup, w_2 that of the stirrer, and w_3 the weight (or volume) of the water displaced by the thermometer; if furthermore s_1 and s_2 are the specific heats, respectively, of the materials of which the inner cup and the stirrer are made,[1] and s_3 the thermal capacity of a quantity of mercury and glass equal in volume to a gram of water;[2] then the thermal capacity of the inner cup is $w_1 \, s_1$; that of the stirrer, $w_2 \, s_2$; that of the thermometer, $w_3 \, s_3$; hence the total thermal capacity of the calorimeter (c) is given by the formula,

$$c = w_1 \, s_1 + w_2 \, s_2 + w_3 \, s_3. \qquad \text{III.}$$

If, for example, the inner cup contains 100 g. of brass, of the specific heat .094, its thermal capacity is

[1] The inner cup and stirrer are usually made of brass (an alloy of copper and zinc), the specific heat of which may be taken as .094.

[2] It will be noted that though the specific heat of mercury (.033) differs greatly from that of glass (0.19), the thermal capacity of *equal volumes* is very nearly the same. Since 1 *cu. cm.* of mercury weighs 13.6 grams, it will require 13.6 × .033, or 0.45 units of heat, to raise it 1°. In the same way, since 1 *cu. cm.* of ordinary glass weighs not far from 2.5 grams, it would require about 2.5 × 0.19, or 0.47 units of heat to raise it 1°. In calculating the thermal capacity of a thermometer, there will be, accordingly, *no appreciable error* in assuming for s_3 a mean value, 0.46.

$100 \times .094$, or 9.4 units; if the stirrer is made of thin brass weighing 2 grams, its thermal capacity is similarly 0.2 units; and if the thermometer displaces 0.9 grams of water, its thermal capacity is (see 2d footnote, page 161) 0.9×0.46, or about 0.4 units. The total thermal capacity of a calorimeter thus constructed would be $9.4 + 0.2 + 0.4 = 10.0$ units.

The first method is apt to give too high results, since the cooling of the water, due to evaporation and other causes, is attributed to contact with the calorimeter.

The second method usually gives too low results, on account of the rapidity with which heat escapes from the calorimeter while empty. If, however, the outer cup becomes heated indirectly by the shot, a portion of this heat may be radiated back to the inner cup when filled with water. It is possible, therefore, that the results may be too great.

The last method generally gives too small a result, because we neglect the heat absorbed by the materials surrounding the inner cup. If, however, only a portion of the inner cup is to be heated, we may easily over-estimate its thermal capacity.

In the latter case, we prefer an experimental determination of thermal capacity; but when the inner cup is made of very thin metal (as is desirable for accurate work), the thermal capacity may be so slight that it cannot be exactly determined by experiment. In such cases, we usually depend upon a calculation based, as in the last method, upon the weights and specific heats of the materials composing the calorimeter.

¶ 92. **Precautions Peculiar to Calorimetry.**— In nearly all experiments in calorimetry two bodies, of known weights and temperatures, are brought together so that by the flow of heat from one to the other (see Experiments 33 and 34) or by the action of one on the other (see Experiments 35–38) a third temperature results. There are, accordingly, many precautions common to these various experiments.

(1) CHEMICAL ACTION.— It is evident that the substances employed should exert no chemical action on the sides of the calorimeter. With strong acids, a glass vessel should generally be employed. Instead of a brass stirrer, one of platinum may be used. In the case of mercury, iron will do even better. A coating of asphaltum is often sufficient to prevent metals from being attacked by acids.

When two substances are placed together in a calorimeter, neither should act chemically upon the other unless the object of the experiment be to measure the heat developed by the reaction. The chemical relations between two substances thus employed must frequently be investigated by a separate experiment.

(2) COMPARISON OF THERMOMETERS.— The general precautions necessary to the accurate observation of temperature have been already considered (¶ 65), and must be observed. In addition to these precautions, certain others are required when *simultaneous* observations of temperature are to be made. In such cases it may be necessary to employ as many thermometers as there are temperatures to be determined; and these thermometers have to be compared with

one already tested by a process of calibration (¶ 68). To do this, the several thermometers are to be placed in boiling water, in ice-water, and in water of at least three intermediate temperatures. A large quantity of water should be used (see (3)), and it must be well stirred in each case. The indications of each thermometer are to be read in turn; then again read *in the inverse order*. There should be regular intervals (let us say 30 seconds each) between the observations. The two readings of each thermometer are to be averaged, and the averages compared. Knowing (from Experiment 25) the corrections for one of the thermometers, we may easily calculate the corrections for the others. For example, if three thermometers, A, B, and C, gave the following readings:

A, 76°.0; B, 75°.7; C, 75°.1; C, 74°.7; B, 74°.5; A, 74°.0;

the average for A would be 75°.0; for B, 75°.1; for C, 74°.9. These averages evidently correspond to the same point of time. We should therefore subtract 0°.1 from the correction of A at 75° to find that of B; and we should add 0°.1 to find that of C.

The object of making observations in the order given above is to eliminate errors due to cooling.

(3) CONSTANT TEMPERATURE.— The difficulty of making accurate observations of temperature at a given point of time increases with the rate of cooling. The use of large masses of water (see (2)) is one of the most general methods of avoiding rapid changes of temperature. In certain experiments in

calorimetry, special devices are frequently employed. When, for instance, one of the temperatures to be observed is in the neighborhood of 100°, a steam-heater may be employed (see Fig. 77, also Fig. 79, ¶ 94). Again, a body may be maintained at 0° by surrounding it with melting ice; or it may be kept indefinitely, without special precautions, at the temperature of the room, provided that the latter be constant.

By the use of devices for maintaining a constant temperature, thermometric observations become greatly simplified. One or more temperatures may be known by definition,—as in the case of ice, or steam at a certain pressure (§ 4). In the absence of cooling, a series of observations for each temperature will not be required, and the temperatures of several bodies at a given point of time may be found from successive observations with the same thermometer. The least constant temperature should be observed nearest the time in question.

(4) EXPOSURE TO THE AIR.— When a body is transferred from a heater or from a refrigerator to a calorimeter, there is always more or less heat gained or lost from exposure to the air. The time of exposure should evidently be made as short as possible. In pouring liquids, a glass funnel may be employed; but the funnel must be warmed to the same temperature as the liquid, otherwise it would take from it more heat than the air. Water may be guided conveniently from a beaker to a calorimeter by a wet glass rod, *abc*, bent as in Figure 76. To

prevent the water from following the side of the beaker, the lip should be greased at the point b.

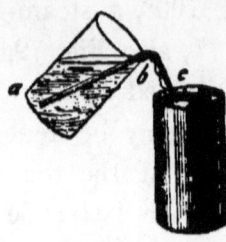

FIG. 76.

The wet stem of a thermometer may also be used as a conductor, and with this advantage, that, since the thermal capacity is easily found (¶ 90, 2) the heat required to raise it to a given temperature may be calculated. We may notice, however, that if the thermometer is immediately afterward placed in the calorimeter, it will give up most if not all of the heat which it has absorbed, and that the remainder may be neglected. Hot shot may be poured directly from a heater suitably shaped (see Fig. 79, ¶ 94) into a calorimeter; but it is safer to use a paper funnel, to prevent the possibility of losing a portion of the shot. Most of the shot should enter the calorimeter without touching the funnel; and the remainder should be in contact with it only for an instant. In this case the heat absorbed by the paper may be neglected. A hot body may also be suspended by a thread, and thus transferred from one place to another.

It is obvious that the calorimeter should be brought as near the heater or refrigerator as is possible without danger that its temperature may be affected by radiation, conduction, or convection from the heater (§ 89). A common pine board makes an excellent shield. In Regnault's apparatus [1] (Fig. 77) the

---

[1] For a fuller description of Regnault's apparatus, see Cooke's Chemical Physics, page 470.

calorimeter (at the left of the figure) can be brought directly under the large steam heater (at the right of the figure). The steam heater rests upon a support, serving to shield the calorimeter from radiation. The support is made hollow, so that it may be kept cool by a current of water. The inner chamber of the heater contains hot air. The temperature within it is observed by means of a thermometer passing through a cork by which the top of the chamber is

Fig. 77.

closed. The bottom of the chamber is closed by a non-conducting slide. By drawing the slide a body suspended by a thread in the hot-air chamber may be lowered directly into the calorimeter. The calorimeter is then immediately removed to a sufficient distance from the heater, so that the resulting temperature may be accurately determined.

By devices similar to those alluded to, the gain

or loss of heat by exposure to the air may be almost indefinitely reduced, but never completely avoided. The student is advised not to attempt any correction for this heat; because a greater error might easily result from applying such a correction than from neglecting it altogether. At the same time, it is well to estimate roughly the quantity of heat gained or lost, with a view to determining what figures of the final result are likely to be affected.

For this purpose two experiments may be made. In one, a body is transferred in the ordinary manner from the heater or from the refrigerator to the calorimeter. In the second experiment, it is passed back and forth let us say 5 times each way, and finally placed in the calorimeter. The body is thus to be exposed to the air in one case about 11 times as long as in the other case, and under similar conditions; so that from the difference in the results we may infer the effect of an ordinary exposure (see ¶ 93, 4).

(5) LOSS OF MATERIAL. — In rapidly pouring a liquid into a calorimeter, or in rapidly lowering a hot solid into a liquid already contained in a calorimeter, there is danger that a portion of the liquid or solid may be lost. It is accordingly desirable to weigh, both before and after each addition to the contents of the calorimeter, not only the calorimeter itself, but also the vessel in which the substance in question was originally contained. The student will do well also to make sure that the space between the inner and outer cups is empty, both before and after

the experiment; for if any of the substance finds its way into this space, its loss will not be apparent from the weighings.

(6) EVAPORATION. — A considerable portion of the heat lost by a liquid when poured into a calorimeter may be caused by evaporation. When once the liquid has been transferred to the calorimeter, all further loss of heat by evaporation should be prevented by immediately corking the inner vessel. It will be assumed that the inner vessel is never uncorked, except when necessary for the purposes of manipulation. Of two liquids, the denser is usually the less volatile, and hence should be heated in preference to the other. For the same reason, a solid should be heated in preference to a liquid. A combustible liquid should, as we have seen (Exp. 30), never be heated directly by a flame, but indirectly by hot water.

(7) TEMPERATURE OF THE ROOM. — The loss of heat which takes place from the gradual cooling of a calorimeter and its contents depends, as we have seen in Experiment 31, upon the difference of temperature between the inner cup and its surroundings. To diminish the loss of heat in question, it has been proposed that the outer cup should be placed in water at the same temperature as the inner cup. More accurate results might be expected from calorimetry if some means were perfected by which we could adjust the temperature of surroundings to the needs of an experiment. In practice, however, the experiment must be adapted to the temperature of the air in

which it is to be performed. When considerable time is required to obtain an equilibrium of temperature (see (8)), it is important that the average temperature within the calorimeter should agree as closely as possible with that of the room. The weights and temperatures of the substances employed in calorimetry, are, therefore, frequently chosen so as to give a final temperature between 20° and 25°.

It is much easier to prevent than to allow for losses of heat by cooling; and it may be stated as a general rule in calorimetry that we must avoid in so far as possible all differences of temperature between bodies under observation and the objects by which they are surrounded.

(8) EQUILIBRIUM OF TEMPERATURE. — It has already been pointed out that to obtain a uniform temperature throughout the inner cup of a calorimeter, the cup should be completely filled. If this is not done, special precautions must be taken to bring its contents into contact with every portion of its surface (see ¶ 85, 2). The necessity of stirring these contents has also been alluded to (¶ 65, 5). When a mixture (like lead shot and water) is of such a nature that an ordinary stirrer cannot be used, the inner cup must be closed water-tight, so that the contents may be shaken. The thermometer should in this case fit tightly into the stopper which closes the inner cup, and should reach into the body of the mixture. Solids, if any be used, should be finely divided, so that there may be no risk of breaking the thermometer.

We prefer, moreover, finely divided solids, on account of the comparative rapidity with which an equilibrium of temperature may be reached, or a process of fusion, solution, or chemical combination completed. When a solid sinks in a fluid (as is generally the case), it is well if it can be warmer than the fluid, on account of the manner in which convection currents are formed; and for the same reason we prefer that the denser of two liquids should have the higher temperature. It is always desirable that the denser of two substances should be poured into the other, so that, as it passes through, as much heat as possible may be communicated from one to the other. The various processes in calorimetry should in general be completed in the shortest possible time, especially when they cannot be conducted at the temperature of the room, since otherwise large losses of heat are apt to occur.

Throughout the processes in question, stirring must be interrupted from time to time, in order that rough observations of temperature may be made. When two successive observations agree, or when they differ by an amount which may be attributed to the regular cooling of the calorimeter (see Exp. 31), the equilibrium of temperature should be complete. The student will do well, however, to make sure that the temperatures at the top and bottom of the calorimeter are the same, before proceeding to make exact observations of the thermometer.

(9) TIMING OBSERVATIONS.—When observations of temperature are taken regularly at intervals of one

or two minutes throughout an experiment, we may infer the time when a given process begins and when it ends; but to avoid errors due to the possible omission of one or more observations, it is well to note the beginning and end of each process in *hours, minutes, and seconds*. In any case, the time should be thus noted, (1st) when all the bodies have been transferred to the calorimeter, and (2d) when, after an equilibrium of temperature has been reached, the resulting temperature is first observed.

(10) SERIES OF TEMPERATURES. — It is well in all cases to make several observations of the final temperature within a calorimeter, in order that the result may not depend upon one alone (see § 51). The series should be made at intervals of one minute, so that, as in ¶ 93 (2), the rate of cooling may be found and allowed for. If the calorimeter contains water only, we may utilize the temperature curves already plotted (see ¶ 93, 1); or if we have determined, as in ¶ 87, the flow of heat from the calorimeter, we may make an allowance for the heat lost as in ¶ 93 (3). In the absence of any previous determination under the same conditions as in the actual experiment, a series of observations of the temperature of the calorimeter will be required.

In the same way, if the temperature of a body is changing perceptibly before it is placed in a calorimeter, it must be determined by a series of observations. The intervals in all such series would naturally be one minute each; but when the temperatures of two or more bodies are to be found, the observations must be taken in turn. When special precautions concerning equilibrium of temperature (see (8)) have to be observed, the student is advised not to attempt observations at intervals of less than one minute. The temperatures of the several bodies concerned are to be reduced in all cases, as in ¶ 93 (1), to the time when they are *first enclosed in the calorimeter*. After this time, losses of heat are to be calculated as above, from the known rate of cooling of the calorimeter.

¶ 93. **Corrections for Cooling.** — (1) GRAPHICAL METHOD. — When a calorimeter contains water only, as in the determination of thermal capacity above (¶ 90, I.) or in parts of various experiments which follow, the temperature at one point of time may be inferred from an observation taken at another

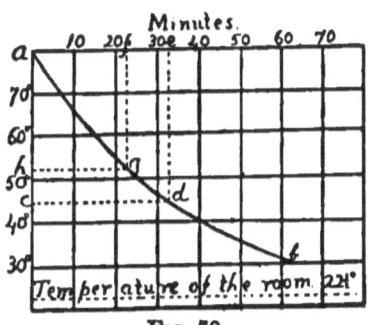

FIG. 78.

point of time by using one of the curves in Fig. 74, ¶ 88. Let *ab* (Fig. 78) be the curve corresponding to the quantity of water which the calorimeter contains, and let *c* be the observed temperature. We first find a point *d* on the curve at the right of *c*, then a point *e* above *d*. Then we measure off a distance *ef* on the scale of minutes corresponding to the length of time during which the calorimeter has been cooling. Then we find a point *g* on the curve below *f*, and finally the temperature *h*, at the left of *g*.

This temperature corresponds in the figure to a time f earlier than e; but by laying off the distance ef to the right of e, we could find, if we chose, the temperature at a later point of time.

A more exact method would be to start with a point c (in Fig. 78), corresponding to a temperature as far above that of the room ($22\frac{1}{2}°$, Fig. 78) as the actual temperature observed was above the observed temperature of the room. The number of degrees included between c and b gives approximately, in any case, the fall of temperature which takes place in an interval of time corresponding to the number of minutes between e and f.

(2) ANALYTICAL METHOD. — When several temperatures have been recorded at regular intervals, we may infer the temperature at a point of time before the beginning or after the end of the series as follows: The observations are first written down in a column, as in the example below; then the temperature of the room is subtracted from each, and the results entered in a second column; then a third column is formed from the differences between each pair of consecutive numbers in the second column; then each number in the third column is divided by the one just below it in the second column, to find what per cent must be added to that number in order to obtain the one above it; these per cents are arranged in a fourth column and averaged; then each number in the third column is divided by the number in the second column just above it, to find what per cent must be subtracted from that number to obtain

the number just below it; the per cents to be subtracted are then arranged in a fifth column and averaged. We may now extend the second column upwards by adding to the first number in it the average per cent from the fourth column, and we may extend it downward by subtracting from the last number the average per cent found in the fifth column. When the second column has been thus extended, the corresponding numbers in the first column may be found by adding in the temperature of the room. The temperature at a time which would come between the observations in the series thus extended may evidently be found by simple interpolation.

For example, when the temperature of the room is 26°, the observations below would be reduced as follows:—

Temperatures Observed.	Temperatures less 26°.	Fall of Temperature.	Per Cent to be Added.	Per Cent to be Subtracted.
66°.0	40°.0			
64°.0	38°.0	2°.0	5.3	5.0
62°.1	36°.1	1°.9	5.3	5.0
60°.5	34°.5	1°.6	4.6	4.4
59°.0	33°.0	1°.5	4.5	4.3
57°.4	31°.4	1°.6	5.1	4.8
56°.0	30°.0	1°.4	4.7	4.5
	Average		4.9	4.7

To extend the second column upwards we add to the first number in it 4.9 per cent of itself. Since 4.9 of 40°.0 is 2°.0, the number above 40°.0 should be 40°.0 + 2°.0, or 42°.0; and since 4.9 per cent of 42°.0 is 2°.1, the next number should be 44°.1, etc.

To extend the second column downwards, we subtract from the last number (30°.0) in it not 4.9 per cent but 4.7 per cent of 30°.0; that is 1°.4; this gives 28°.6; and subtracting from this 4.7 per cent of itself, or 1°.3, we find 27°.3 for the number following, etc.

Adding 26° to the new numbers in the second column, we infer, finally, that the temperatures preceding 66°.0 in the first column should be 68°.0 and 70°.1, while those following 56°.0 should be 54°.6 and 53°.3, etc.

Let us suppose that the temperatures were observed at intervals of one minute; then to represent the temperature for instance 1.5 minutes before the first recorded observation, we should take a number half-way between 68°.0 and 70°.1, or 69°.0 nearly. If, however, the intervals between observations were two minutes each, then 1.5 minutes would be three fourths of one interval, and we should add to 66° three fourths of the difference (2°) between it and the next temperature above it in the series to find the temperature (67°.5) in question.

The discovery of various methods by which the calculations described above may be shortened, especially by the use of logarithms, may be left to the ingenuity of the student. The method here described is important, as an illustration of the fact that when a body is steadily cooling its temperature falls, not a given amount in each minute, but a certain *per cent* (approximately) of the number of degrees which lie between it and the temperature of the room (see ¶ 86, 1).

The accuracy with which a series of observations

may be extended by analytical methods evidently grows less as the number of new terms increases. It may be said in general that the new terms should not be more numerous than those obtained by actual observation.

(3) HEAT LOST BY COOLING. — We must distinguish between the rate of cooling of a calorimeter and the number of units of heat lost by it. The latter may be found without knowing the nature of the mixture which the calorimeter contains, provided that the inner cup is completely filled by the mixture, or filled to a known depth; for we have only to refer to the results already found with water at the same depth in Experiment 31.

If, for example, a calorimeter, nearly filled with a mixture of lead shot and water, has been cooling for ten minutes at an average temperature about 20° above that of the room, we reason that since at a temperature 1° above that of the room it was found (¶ 87) to lose 1.8 units of heat per minute, at a temperature 20° above that of the room it would lose 20 times 1.8, or 36 units per minute; that is, 360 units in ten minutes. If, therefore, the first accurate observation of temperature was taken ten minutes after the introduction of the mixture, we should add 360 units to the amount of heat apparently given out by the hot body, or if more convenient we may subtract 360 units from the quantity of heat apparently absorbed by the cool body (see ¶ 98).

(4) METHOD OF MULTIPLICATION. — When two experiments are made, in one of which a body is exposed,

let us say, 11 times as long or 11 times as often to the air as in the other experiment, in which we give it the ordinary exposure, the difference between the results obtained in the two cases should correspond to the effect of 11 less 1, or 10 ordinary exposures. Hence, if this difference be divided by 10, we may estimate roughly the correction to be applied to the result obtained with the ordinary exposure.

If, for example, the thermal capacity of a calorimeter is found to be 10.1 units when warm water is poured into it directly, and 11.1 units if the water is first poured back and forth five times each way, then the effect of cooling due to 10 transfers is 11.1–10.1, or 1 unit in the result; and the effect of a single transfer is about 0.1 unit. The true thermal capacity is, therefore, about 10.1–0.1, or 10.0 units. If the cooling due to transferring a substance from one place to another is thought to affect the figure in the tenths' place, as in the example, it is evident that the hundredths will not be significant (see § 55).

EXPERIMENT XXXIII.

SPECIFIC HEAT OF SOLIDS.

¶ 94. **Determination of the Specific Heat of a Solid by the Method of Mixture.** — I. A quantity of lead shot sufficient to half fill the calorimeter (Fig. 70, ¶ 85) is first weighed, then put into a steam heater (Fig. 79), and covered by a cork. A thermometer, passing through the cork into the midst of the shot, is allowed to remain there until it ceases to rise. Meanwhile the temperature within the calorimeter is determined by a second thermometer (¶ 92, 2). The calorimeter is then weighed, and a vessel containing a mixture of ice and water is also weighed. This vessel should be provided with a strainer, so that water may be poured from it without danger of particles of ice following the stream.

FIG. 79.

The ice and water should be thoroughly stirred just before the experiment, to secure a uniform temperature of 0°. The time should now be noted (¶ 92, 9).

The thermometer and corks are then removed from the heater, and the shot is poured as rapidly as possible (¶ 92, 4) into the calorimeter. Immediately ice-cold water is added, — the quantity being nearly sufficient to fill the calorimeter. A thermometer is then pushed cautiously into the middle of the shot through a small stopper, closing the inner cup watertight (¶ 92, 8). The large cork cover (Fig. 71, ¶ 85) may then be added, and the time again recorded. The mixture must now be carefully shaken. The temperature indicated by the thermometer is to be noted at intervals of 1 minute, until it begins to fall steadily (¶ 92, 8 and 10). Then the calorimeter is re-weighed with its contents; and the vessel originally containing the water is also weighed (¶ 92, 5).

II. Instead of finding the temperature of the shot in the heater, as in I., we may determine it by a series of observations in the calorimeter, before the ice-water is added (¶ 92, 10). It is necessary in this case to cork the inner cup, and to shake the shot between the observations of temperature (¶ 92, 8), in order that there may be a uniform temperature not only in the shot, but also in the inner cup, the thermal capacity of which must be considered. The ice-water is finally added, and the temperature of the mixture determined as before.

III. Instead of pouring the hot shot first into the calorimeter, we may begin by introducing ice-water. In this case the proper quantity of water must be determined beforehand. It will probably be found that the water should fill the calorimeter about half-full. In other respects this method is the same as I.

IV. Instead of assuming that the temperature of the water is the same as that within the vessel originally containing it (that is, $0°$), we may find its temperature after it has been transferred to the calorimeter. In this method, however, as in the second method, the thermal capacity of the cup must be considered. To avoid the necessity of making a separate series of observations (¶ 92, 10) between which the water in the calorimeter must be shaken up (¶ 92, 8), it is customary to use water at the temperature of the room. In this case, the mixture will be above the temperature of the room; hence its rate of cooling must be allowed for (¶ 93).

V. Other methods of determining specific heat

may easily be devised, depending upon the use of hot water and cold shot. We have in fact already made use of such a method in finding the thermal capacity of a calorimeter (¶ 90, I.). On account, however, of the practical difficulties arising from evaporation (¶ 92, 6), the high temperature of the mixture (¶ 92, 7), and the small change of temperature produced, these methods are generally avoided. The principal use which can be made of them is as a check (§ 45) upon results obtained in the ordinary manner.

The student may observe that in the second method the shot falls suddenly in temperature, on account of the heat which it gives up to the calorimeter. This heat is subsequently restored to the mixture when the calorimeter is cooled to its original temperature; hence in the first method no account need be taken of the thermal capacity of the calorimeter. Again in the fourth method, the cold water may at first rise rapidly in temperature on account of the heat imparted to it from the calorimeter, but this heat is restored to the calorimeter when it is again raised by the mixture to its original temperature; hence in the third method no account need be taken of the thermal capacity of the calorimeter.

Instead of lead shot, copper or iron rivets may be employed with very slight modifications of the experiment. In the case however of solids which are soluble in water, we must substitute for water some other liquid of known specific heat in which the solids are insoluble (¶ 92, 1). The student may be guided in his choice of methods by obvious considerations of practical convenience as well as by the principles explained below in ¶ 95; but he should make at least one determination of specific heat of a solid by the method of mixture and reduce it as will be explained in ¶ 98.

¶ 95. **Comparison of Methods for the Determination of Specific Heat.** — The principal difficulty in the first method (¶ 94, I.), for the determination of specific heat, is to avoid a great loss of heat while the shot is being transferred from the heater to the calorimeter.

In the second method (¶ 94, II.) there is no opportunity for a loss of heat on the part of the shot, since its temperature is determined by a series of observations within the calorimeter, from which its temperature at any point of time may be found (¶ 93, 2). The principal objection to the second method is the difficulty of determining accurately a series of temperatures in which rapid changes take place; and the necessity of allowing for the thermal capacity of the calorimeter, which is always a more or less uncertain quantity, and bears a considerable proportion to the thermal capacity of the shot.

The third method (¶ 94, III.) has the same practical advantages and disadvantages as the first.

The fourth method (¶ 94, IV.) is the one commonly employed for the determination of specific heat. Since the temperature of the water is found when within the calorimeter, there is no opportunity (as in the other methods) for heat to be imparted to it in the act of pouring. There is however difficulty, as in the second method, in determining accurately a temperature which is changing (¶ 92, 3), and still further difficulty in maintaining a uniform temperature throughout the calorimeter with a quantity of water which only half fills it (¶ 92, 8). When the latter difficulty is avoided by using water at the temperature of the room, the mixture must have a temperature considerably above that of the room, and one therefore which is hard to determine (¶ 92, 3). The thermal capacity of the calorimeter must also, as in the second method, be taken into account.

By comparing the results of the first and second methods, we are able to estimate the effect of the heat lost in pouring the shot into the calorimeter (see also ¶ 93, 4), and by comparing results of the third and fourth methods, we are able to estimate the effect of the heat absorbed by the ice-cold water when it is poured from one vessel to another. This will be found to be small in comparison with the heat lost by shot at 100° under similar circumstances. The second method, in which the latter is eliminated, is therefore preferable to the fourth. In the first and third methods, the heat lost by the shot is partly offset by that imparted to the water. Since the former is greater than the latter, the third method is preferable to the first; because the longer exposure of the water may compensate for the more rapid cooling of the shot. The choice between the second and third methods will depend largely upon the comparative accuracy with which we can determine the heat given out by the calorimeter (¶ 87) and the heat lost by the shot (¶ 93, 4). The advantages of using in any case hot shot and cold water have been already stated (¶ 94, V.).

EXPERIMENT XXXIV.

SPECIFIC HEAT OF LIQUIDS.

¶ 96. Determination of the Specific Heat of a Liquid by the Method of Mixture. — The specific heat of a liquid may be determined either by mixing it mechanically with water, or by bringing it in contact with a solid of known specific heat. The first method is the more direct, but cannot be employed with liquids which unite chemically with water, unless we know the amount of heat given out or absorbed by the reaction (see ¶ 92, 1). Before deciding which method we shall employ, we therefore mix together the contents of two test-tubes, each at the temperature of the room, one containing water, the other the liquid in question. If no change of temperature is observed, the first method is adopted. If the temperature rises or falls, we must either make a separate experiment to determine accurately the amount of this rise or fall (see Exp. 35), or else adopt the indirect method, using a solid instead of water.

I. The determination of the specific heat of an insoluble liquid by the method of mixture does not differ essentially from the case of a solid. A heavy oil may for instance be heated by the same apparatus (Fig. 79, ¶ 94) employed for the shot, and mixed with ice-cold water, according to either of the methods described (¶ 94). Instead of shaking the mixture, a brass fan or stirrer (Fig. 50, ¶ 65) may be employed.

The objections to mixing hot water with a cold liquid are not nearly as strong as in the case of solids (¶ 94, V.); for though most liquids have a specific heat less than that of water, the differences are very much less. By pouring a comparatively small quantity of water at a temperature not exceeding 40° or 50° into a liquid at 0° a mixture may be had not far from the temperature of the room. With liquids less dense than water this method is generally to be preferred (see ¶ 92, 6 and 8). The results may be reduced by the appropriate formula from ¶ 98.

Attention has already been drawn (¶ 92, 1) to precautions against chemical action in the case of corrosive liquids, and in the case of volatile liquids against evaporation (¶ 92, 6) and combustion (¶ 83).

II. In the case of liquids which mix with water, the ordinary methods of mixture cannot generally be employed, on account of the heat absorbed or developed by solution or combination. It is necessary to find some substance, of known specific heat, upon which such a liquid exerts no thermal action. This substance is then mixed with the liquid by either of the methods of ¶ 94. The data necessary for finding the specific heat of the liquid are as usual the weight of the two substances in question, the temperature of each before the experiment, and the resulting temperature of the mixture.

The lead shot already employed (¶ 94) may be used to determine in this way the specific heat of alcohol, glycerine, saline solutions, etc. For corrosive liquids, like nitric acid, glass beads (of specific heat about 0.19) may be similarly employed (see general formula, ¶ 98). Evidently this indirect method is more general than the ordinary method of mixture, since it can be applied to all liquids, whether soluble or insoluble in water. It has the advantage of eliminating almost completely the heat lost by the hot body between the heater and the calorimeter, since this loss is practically the same in the case of water as in the case of other liquids with which a comparison is made.

¶ 97. **Peculiar Devices employed in Calorimetry.** — In the method of mixture (Exps. 33 and 34) a thermal equilibrium between two or more substances is established by bringing them in contact. It is not, however, necessary that the two bodies should touch each other. The difficulties which arise from the mutual action of two substances may often be avoided by surrounding one of them with an envelope, through which, by the conduction of heat, an equalization of temperature takes place. If, for instance, a hot liquid contained in a glass bulb be surrounded by cold water, a certain quantity of heat will be given out. Having found by a separate experiment how much heat is derived from the bulb alone, we may calculate the specific heat of the liquid in the ordinary manner, that is, from the weights and changes of temperature involved (see general formula, ¶ 98).

The liquid in question may be contained in an ordinary thermometer bulb. In this case its change of temperature may be inferred very accurately from its contraction, as shown by the fall of a column of liquid in the stem of the thermometer. It is necessary, of course, to make a careful comparison of a thermometer containing an unknown liquid with an ordinary mercurial thermometer (see ¶ 92, 2). This method has obvious advantages in the case of costly liquids.

On the other hand, when the supply of a fluid is unlimited, it is frequently advantageous to use an envelope in the form of a spiral tube, or coil, through which the fluid in question may be passed in a continuous stream. We are thus enabled to bring a great volume of the fluid in thermal equilibrium with a small volume of water. This device is exceedingly important in the case of gases, since it would be otherwise impossible to bring enough gas in thermal equilibrium with a given quantity of water to affect the temperature of the water by a measurable amount.

The weight of the gas employed is not measured directly, but is determined from its density (see ¶¶ 44, 46) and from the volume employed. The volume is indicated by a gas-meter (ab, Fig. 80) through which the gas is first passed. The gas is then raised to the temperature of 100° by passing it through a steam jacket, bd. Then it circulates through a coiled tube surrounded with water, and escapes from an orifice where its final temperature can be observed. From the thermal capacity and rise of temperature of the calorimeter, we may calculate the quantity of heat given out by a known quantity of gas in falling through a known number of degrees, and hence the specific heat of the gas. It is found that the specific heat of air at the constant pressure of one atmosphere is about 0.238, or a

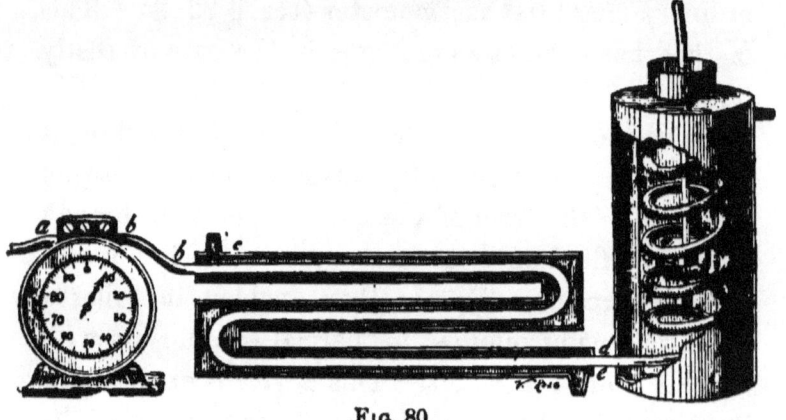

FIG. 80.

little less than one fourth that of an equal weight of water.

A much more difficult task consists in the determination of the specific heat of a gas when confined to a constant volume. The following method is suggested. It depends upon the fact that a given electric current passing for a given time through a given conductor generates in that conductor a given quantity of heat. This quantity may be found by experiment (see Exp. 86), or calculated by the principles of § 136. Let us suppose that a known quantity of heat is thus suddenly generated within a closed flask (Fig. 81); and that the increased pressure of the air is measured, as in ¶ 80, by the rise of mercury in an open tube. Then the average temperature of the air within the flask can be calculated (see § 76).

We may therefore find the thermal capacity of a known volume or of a known weight, and hence the specific heat in question (about .169).

It is found that the thermal capacity of a cubic metre of air is about 219 units at 0° and 76 cm. when prevented from expanding, as against 308 units when free to expand under a constant pressure. The thermal capacity of an *equal volume* of oxygen, of nitrogen, or of hydrogen is very nearly the same as that of air under similar conditions.

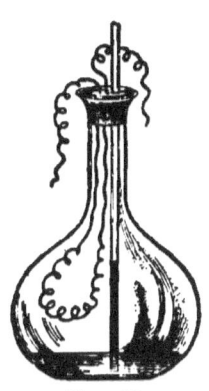

FIG. 81.

Instead of using an electrical current to generate heat (as illustrated in Fig. 81), we may employ various other agents, as for instance the combustion, the solidification, the fusion, the condensation, or the vaporization of a known weight of a given substance, or the conversion through friction of a given amount of work into heat (see Exp. 70). If, for example, the combustion of a gram of coal heats a kilogram of water 8°, and a kilogram of petroleum 16°; or if 100 grams of ice cool these liquids 8° and 16° respectively; the specific heats must be to each other as 2 to 1. The same inference would be drawn if the same quantity (100 grams) of steam which heats 1 kilogram of water 54° were found to heat 2 kilograms of petroleum by the same amount. The specific heats of different substances are to each other, in general, inversely as the changes of temperature produced by a given cause, and also inversely as the weights affected. The determination of specific heat is evidently capable of as many modifications as there are different methods by which a definite quantity of heat may be generated or absorbed.

Instead of using the pressure of air to measure its temperature, we may also employ its expansion (§ 80) as in the air thermometer (¶ 74). The specific heat of air under a constant pressure might obviously be determined by an apparatus similar to that represented in Fig. 81; hence, conversely, if this specific heat is known, we may measure quantities of heat by the expansion which they produce in air at a given pressure. It does not (as one might think) make any difference theoretically *how much* air is heated; because an increase in the quantity of air will be offset by a decrease in the temperature to which it will be raised by a given amount of heat; and for the same reason it is indifferent whether a small portion of the air is heated a great deal, or whether a considerable portion is heated by a proportionately small amount. In this method of estimating heat it is not necessary to wait for an equilibrium of temperature. We hasten in fact to make our observations before an equilibrium is reached, so as to avoid loss of heat by contact of the air with the sides of the vessel in which it is contained. It has been calculated that one unit of heat should in all cases cause in a body of air at 76 *cm.* pressure an expansion of about 12 cubic centimetres. Since an expansion of less than 1 cubic millimetre is easily detected, we have, in the air thermometer, a very delicate means of measuring small quantities of heat.[1]

Instead of air, we may use any other fluid which has a regular rate of expansion to determine quantities of heat. The principle above explained has been applied by Favre and Silbermann in the construction of their mercury calorimeter.[2] This is essentially a thermometer with a huge bulb. If even a small quantity of hot liquid be introduced into a cavity in this bulb, there will be a perceptible expansion of the mercury, by which we may measure the heat given out by the liquid in question; for it has been found that 1 unit of heat always causes in a body of mercury an expansion of about 4 cubic millimetres.

There are various other definite effects produced by a given quantity of heat, any one of which might theoretically be applied to the purposes of calorimetry. The only application of practical importance depends, however, upon the heat required for the fusion of ice (see Experiment 36). A rough form of ice calorimeter consists of a block of ice (Fig. 82) with a small cavity in which a hot body may be

Fig. 82.

---

[1] The air thermometer has been used in the Jefferson Physical Laboratory to measure minute quantities of heat generated in a carbon fibre by telephone currents.

[2] See Ganot's Physics, § 463.

placed. A second block may be used as a cover. The water formed by the liquefaction of ice is gathered by a sponge, and weighed by the usual method of difference. Since one unit of heat melts one-eightieth of a gram of ice, the quantity of heat given out by the body in falling to a temperature of 0° can easily be calculated. In Bunsen's ice calorimeter, the quantity of ice melted is estimated by the change *in volume* of a mixture of ice and water.

¶ 98. **Calculation of Specific Heat in the Method of Mixture.** — If w_1 is the weight of the body, the specific heat of which (s_1) is to be determined, and t_1 the temperature of this body, reduced to the time of mixing; if w_2 is the weight of the body the specific heat (s_2) of which is known, and if t_2 is its temperature, also reduced to the time of mixing; if c is the thermal capacity of the calorimeter, t_3 its original temperature and, t the temperature of the mixture; then if q is the quantity of heat lost by cooling, that is, absorbed by the air, etc., we have, by the principle of § 90, the general formula,

$$w_1 s_1 (t - t_1) + w_2 s_2 (t - t_2) + c (t - t_3) + q = 0.$$

From this formula we may obtain the solution of all problems in the determination of specific heat by the method of mixture.

In addition to s_2, c, and q (which are known, or may be calculated), we require at least five data for a determination of specific heat; namely, the two weights employed, w_1 and w_2, the two corresponding temperatures, t_1 and t_2, also the temperature, t, of the mixture. The original temperature, t_s, of the calorimeter must also be determined, unless by the nature of the experiment it is known to agree with one of the other temperatures.

When water is used $s_2 = 1$; hence we have, if the water used is colder than the mixture,

$$s_1 = \frac{w_2(t-t_2) + c(t-t_3) + q}{w_1(t_1-t)}; \qquad \text{I.}$$

or if the water is warmer than the mixture,

$$s_1 = w_2 \frac{(t_2-t) - c(t-t_3) - q}{w_1(t-t_1)}. \qquad \text{II.}$$

If the temperature of the water is taken in the calorimeter, so that $t_2 = t_3$, we may combine the terms in the numerator, so that for cold water,

$$s_1 = \frac{(w_2+c)(t-t_2) + q}{w_1(t_1-t)}; \qquad \text{III.}$$

or for hot water,

$$s_1 = \frac{(w_2+c)(t_2-t) - q}{w_1(t-t_1)}. \qquad \text{IV.}$$

If the original temperature of the calorimeter is the same as that of the mixture, the terms $c(t-t_3)$ and $c(t_s-t)$ disappear from I. and II. respectively; hence, for cold water,

$$s_1 = \frac{w_2(t-t_2) + q}{w_1(t_1-t)}; \qquad \text{V.}$$

and for hot water,

$$s_1 = \frac{w_2(t_2-t) - q}{w_1(t-t_1)}. \qquad \text{VI.}$$

If, finally, the temperature of the mixture is the same as that of the room, there is no loss of heat by cooling (§ 89), that is, $q = 0$; hence the term q disappears from all the formulæ. We have therefore in the simplest possible case, when the calorimeter is at the temperature of the room both before and after the experiment, if cold water is used,

$$s = \frac{w_2(t-t_2)}{w_1(t_1-t)}; \qquad \text{VII.}$$

and if hot water is used,

$$s = \frac{w_2(t_2-t)}{w_1(t-t_1)}. \qquad \text{VIII.}$$

The calculation of the thermal capacity of the calorimeter (c) is explained in ¶¶ 86 and 91; that of the heat lost (q) in ¶ 93, 3. The correction of the temperatures t_1 and t_2 to the time of mixing may be done either by graphical or by analytical methods (¶ 93, 1 and 2).

EXPERIMENT XXXV.

HEAT OF SOLUTION.

¶ 99. **Determination of Latent Heat of Solution.** — When a solid dissolves in a liquid, or when two liquids mix together, there is almost always a rise or fall of temperature. This is due probably to a a molecular re-arrangement which takes place. The object of this experiment is to find how much heat is given out or absorbed, as the case may be, by one gram of a given substance when mixed with or dissolved in water.

I. LIQUIDS. — When equal volumes of alcohol and water are mixed together (see ¶ 96) a rise of temperature may be observed. To measure this rise accurately, a calorimeter is to be weighed empty, and re-weighed with a quantity of alcohol which fills it half-full, and which is at a temperature, accurately observed, not far from that of the room. An equal volume of water, heated or cooled if necessary so as to have exactly the same temperature, is then mixed with the alcohol in the calorimeter, and the resulting temperature accurately determined by a series of observations (¶ 92, 10). The weight of water is also to be found (see ¶ 92, 5). If the thermal capacity of the calorimeter and the specific heat of the liquid are both known, the latent heat of solution may be calculated by formula II., ¶ 100.

It is better, however, to repeat the experiment with water at a much lower temperature, which must be determined (see ¶ 92, 10) by a series of observations. The object aimed at is to offset in this way the heat due to mixture. When alcohol in a calorimeter at the temperature of the room is mixed with an equal volume of water, which is cooler than it by the right number of degrees, scarcely any rise or fall of temperature will be observed in the calorimeter. In this case a single observation will suffice.

Let us suppose, for example, that equal volumes of alcohol and water rise 8° when mixed at the same temperature, but that if the water is 9° cooler than the alcohol, the rise is 2°. Then since 9° in the water makes a difference of 8° — 2°, or 6°, in the mixture, 12° in the water would make a difference of 8° in the mixture. It follows that the alcohol could be mixed with an equal volume of water 12° below it in temperature without being warmed or cooled by the process.

It would be well to test the accuracy of such a conclusion by a third experiment. When the desired difference of temperature has been found, either by experiment or by calculation, the latent heat of mixing is easily computed. We multiply the weight of water by its rise of temperature to find the number of units of heat received, and divide by the weight of alcohol to find the amount given out by one gram; or we may use formula III., ¶ 100.

The experiment may be varied by using different liquids, or by mixing a given liquid with water in different proportions.

II. SOLIDS. — When ammonic nitrate is dissolved in water a fall of temperature is observed. The amount of this fall may be determined as in the case of alcohol; but in order that the solid may be readily dissolved, it is better to use only one part of the salt in nine of water. To ensure rapid solution, the salt should be pulverized. In the first experiment the salt, the water, and the calorimeter should all start at the temperature of the room. The fall of temperature of the water may require a thermometer divided into tenths of degrees for its accurate determination. The use of a stirrer is very important (¶ 65, 5).

The experiment may now be repeated with water somewhat warmer than before, with a view to making the resulting temperature agree with that of the room. The water should, however, be placed first in the calorimeter, in order that the temperature of the latter may be accurately determined. A series of observations must be taken (¶ 92, 10). The salt is finally added, and the fall of temperature accurately measured. If the water has been heated too much or too little, the experiment may be repeated until the mixture agrees in temperature with the room; or the desired temperature of the water may be calculated by the same process of reasoning as was employed in I. In calculating the latent heat of solution by this method, the thermal capacity of the calorimeter must be taken into account, since part of the heat absorbed by the salt is supplied by the calorimeter. In other respects the reduction is the same as in I. (see also formula IV., ¶ 100).

If, for instance, 10 grams of salt cool 90 grams of water contained in a calorimeter with a thermal capacity equal to 10 units, from 22° to 20°, that is 2°, we have $(90 + 10) \times 2 = 200$ units of heat given out. Since 10 grams of the salt absorb 200 units, each gram must require 20 units of heat; hence the latent heat of solution is 20. The latent heat in question varies slightly according to the strength of the solution formed.

¶ 100. **Calculation of the Latent Heat of Solution.** — If w_1 is the weight of the substance whose latent heat of solution, l_1, is to be determined, s_1 its specific heat, and t_1 its original temperature; if w_2 is the weight of the solvent, s_2 its specific heat, and t_2 its original temperature; if c is the thermal capacity, t_3 the original and t the final temperature of the calorimeter (hence also of the mixture), then the quantities of heat absorbed are, (1) $w_1 \, s_1 \, (t - t_1)$ in raising the temperature of the substance dissolved; (2) $w_2 \, s_2 \, (t - t_2)$ in raising the temperature of the solvent; and (3) $c \, (t - t_3)$ in raising the temperature of the calorimeter and (4) $w_1 \, l_1$ in the act of solution. Hence, by the principle of § 90,

$$w_1 s_1 (t - t_1) + w_2 s_2 (t - t_2) + c(t - t_3) + w_1 l_1 = 0, \quad \text{I.}$$

neglecting the heat lost by cooling.

This gives for the latent heat of mixing with water, which we consider positive if heat is absorbed, but negative if (as is usually the case when two liquids are mixed) heat is given out,[1] since $s_2 = 1$, and since t_1 and t_3 are the same (the temperature of the liquid being determined in the calorimeter),

$$l_1 = - \frac{(w_1 \, s_1 + c)(t - t_1) + w_2 (t - t_2)}{w_1}. \quad \text{II.}$$

If the experiment is varied so that $t = t_1$ then we have simply

$$l_1 = - \frac{w_2 (t - t_2)}{w_1}. \quad \text{III.}$$

If, however, the temperature of the water is found within the calorimeter, so that $t_2 = t_3$, the substance

[1] The same formula may be used to determine the heat of combination, only that the sign must be reversed (see ¶ 106).

dissolved being as before unchanged in temperature, we have for the latent heat of solution, which we call positive when heat is absorbed, the formula

$$l_1 = \frac{(w_2 + c)(t_2 - t)}{w_1}. \qquad \text{IV.}$$

EXPERIMENT XXXVI.

LATENT HEAT OF LIQUEFACTION.

¶ 101. **Determination of the Latent Heat of Water.** — Latent heats of liquefaction are determined in essentially the same manner as latent heats of solution (Exp. 35, II.). Instead, however, of dissolving a solid in a fluid, the solid is simply melted by the fluid. Knowing the weights, specific heats, and changes of temperature of the substances in question, we may calculate by the general formula (¶ 100, I.) the heat required to melt one gram of the solid; or, in other words, its latent heat of liquefaction.

It is evident that the liquid must exert no solvent action on the solid, otherwise we should have to allow for heat of solution (see Exp. 35). It is also necessary that the mixture be at a higher temperature than the solid, else the solid will not melt. It is well that the solid should start at its melting-point, since otherwise we must allow for the heat necessary to raise it to the temperature in question. A considerable time must generally be allowed for the process of melting; to shorten this time as much as possible, the mixture should be vigorously stirred. Observations of temperature should be taken from time to time (¶ 92, 8) during the process.

When ice is the solid employed, difficulty will be found in obtaining sufficiently small pieces free from water. The ice should be cracked into fragments weighing a few grams each, which are then to be wrapped up in cotton-waste and weighed. Any moisture formed by the melting of the ice should be absorbed by the waste.

The calorimeter is weighed empty, and re-weighed when about half-full of warm water. The temperature of the water should be about 50°, and is determined by a series of observations (¶ 92, 10); then ice is added until the calorimeter is nearly full. The ice should be handled by means of a portion of the cotton waste which surrounds it, and each fragment should be wiped as dry as possible before placing it in the calorimeter. The time occupied by this process and by the fusion of the ice should be noted (¶ 92, 9). The resulting temperature of the water must be accurately determined. The quantity of ice used should be found both by re-weighing the cotton waste and by re-weighing the calorimeter (¶ 92, 5).

¶ 102. **Calculation of the Latent Heat of Water.** — If w_1 is the weight of ice employed, t_1 its original temperature (that is, 0°) and s_1 its specific heat in the liquid state (that is, 1); if w_2 is the weight of water employed, t_2 its temperature reduced to the time of mixing (¶ 93), and s_2 its specific heat (that is 1); if c is the thermal capacity of the calorimeter calculated as in ¶ 91, t_3 its original temperature (the same as t_2), and t the temperature of the mixture; we have, substituting these values in formula II., ¶ 100, —

$$l_1 = \frac{(w_2 + c)(t_2 - t) - w_1 t}{w_1}.$$

From the numerator of this fraction should be subtracted a correction expressing the number of units of heat lost by the warm water while the ice is being melted. Since the water begins at a temperature t_2, and ends at a temperature t, its average temperature is $\frac{1}{2}(t_2 + t)$, nearly. Subtracting the temperature of the room, we have, approximately, the average excess of temperature. Multiplying as in ¶ 93 (3), by the number of minutes required to melt the ice, and also by the heat lost per minute when the temperature is 1° above that of the room (see ¶ 87), we have the correction in question. Evidently, if the average temperature of the water is the same as that of the room, no correction for cooling need be made.

The truth of the formula for the latent heat of water may be seen by the following considerations: Since w_2 grams of water and the equivalent of c grams of water (in the brass and other materials composing the calorimeter) are cooled from $t_2°$ to $t°$, the heat lost by the hot bodies amounts to $(w_2 + c) \times (t_2 - t)$ units. Subtracting from this the correction for cooling, we have a remainder which must represent the heat absorbed by the cold bodies; that is, the ice and the water formed by its liquefaction. Now w_1 grams of ice form w_1 grams of water at 0°;

and to raise this to $t°$ requires $w_1 \times t$ units of heat. Subtracting this from the previous remainder, we have, therefore, the heat required to melt w_1 grams of ice. Finally, dividing by w_1, we have the heat required to melt 1 gram, or the latent heat in question.

EXPERIMENT XXXVII.

LATENT HEAT OF VAPORIZATION.

¶ 103. **Determination of the Latent Heat of Steam.** — There are many points of resemblance between the determination of the latent heat of vaporization and that of the latent heat of liquefaction (Exp. 36). Instead of melting a solid in a liquid, a vapor is condensed in a liquid. From the weights, specific heats, and changes of temperature in question, latent heats of vaporization may be calculated by the *same general formula* (¶ 100, I.) as latent heats of liquefaction.

The vapor must evidently have no chemical affinity for the liquid. The liquid must be at lower temperature than the vapor, in order that the latter may be condensed. The vapor should start as nearly as possible at its temperature of condensation, otherwise an allowance must be made for the heat given out in reaching this temperature. Care must, however, be taken that the vapor is freed from particles of liquid formed by its condensation, before it passes into the calorimeter.

¶ 103.] LATENT HEAT OF VAPORIZATION. 203

When steam is used, it is passed from a generator (*a*, Fig. 83) through a trap (*b*), where nearly all its moisture is deposited. It will be seen in the diagram that the exit tube is completely surrounded, either by steam or by cork, until it reaches the calorimeter. If, therefore, this tube is well heated by a current of steam before the experiment, there is no reason why any condensation should take place within it.

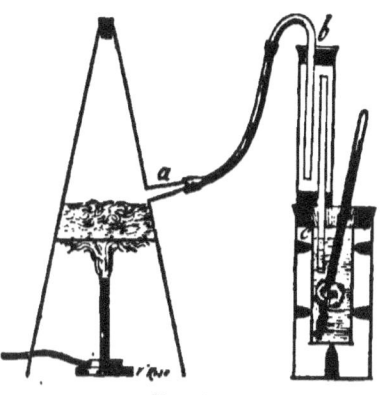

Fig. 83.

The calorimeter is weighed when empty, and reweighed with a quantity of water sufficient nearly to fill the inner cup, and as cold as possible. The temperature of this water is determined by a series of observations at intervals of one minute (¶ 92, 10); then the current of steam issuing from the trap is turned suddenly into the water. The water is stirred vigorously by twisting the stem of a thermometer to which a stirrer is attached. When the temperature of the water has risen as much above that of the room as it was below it before the admission of steam, the trap is taken away from the calorimeter, and the resulting temperature determined by another series of observations. The time used in heating the water to the required temperature should be as small as possible, to avoid errors due to gain or loss of heat; but if the *average* temperature agrees with that of the room, no correction for cooling need be applied (see ¶ 102). The weight of steam condensed is found by re-weighing the calorimeter, and the temperature of this steam determined by an observation of the barometer (see ¶ 69, II.).

¶ 104. **Calculation of the Latent Heat of Steam.** — If w_1 is the weight of steam condensed, s_1 the specific heat of the liquid formed by its condensation (that is, 1),[1] and t_1 its original temperature (let us say 100°, but see Table 14); if w_2 is the weight of water, s_2 its specific heat (that is, 1) and t_2 its original temperature; if c is the thermal capacity of the calorimeter, t_3 its original temperature (the same as t_2), and t the temperature of the mixture; we have, substituting these values in the general formula (¶ 100, I.), —

$$l_1 = \frac{(w_2 + c)\ (t - t_2) - w_1\ (100 - t)}{w_1}.$$

To the numerator of this fraction should be added the heat (if any) lost in cooling, since this is also at the expense of the steam.

The formula may also be established by a process of reasoning similar to that used in ¶ 102. To raise the equivalent of $w_2 + c$ grams of water $(t - t_2)$ degrees requires $(w_2 + c) \times (t - t_2)$ units of heat. Part of this was furnished by the w_1 grams of water at 100° (nearly) in cooling to $t°$. This part is clearly $w_1\ (100 - t)$. Subtracting this from the total heat

---

[1] The specific heat of water varies from 1.000 at 0° to 1.013 at 100°, having a mean value of about 1.005.

received by the water, we have that given up to it by w_1 grams of steam in the act of condensation; hence, dividing by w_1, we have the heat given out by one gram of steam at 100° when condensed into water at 100°; that is, the latent heat in question.

EXPERIMENT XXXVIII.

HEAT OF COMBINATION.

¶ 105. **Determination of Heats of Combination.** — The same method, essentially, is employed for the determination of heats of combination as for heats of solution (Experiment 35); the only difference being that the solvent has a chemical affinity for the substance dissolved. From the weights, specific heats, and changes of temperature of the materials involved, the heat of combination may be calculated by the general formula (¶ 100, I.). Heats of combination are, however, called positive when the result of mixture is to raise the temperature of the constituents.

(1) ZINC AND NITRIC ACID. — A gram of pure zinc filings is to be dissolved in at least fifty times its weight of dilute nitric acid. The student should determine by a preliminary experiment what strength of acid may be required to ensure rapid solution without danger of accident from excessive effervescence. This will depend largely upon the fineness of the zinc. When "zinc dust" is used, very dilute acids must be employed. The zinc dust should be

poured into the acid, not the acid on the zinc dust. The inner cup of the calorimeter (Fig. 71, ¶ 85) should be replaced by one of glass (¶ 92, 1), the thermal capacity of which must be calculated as in ¶ 91. The glass cup is then nearly filled (¶ 92, 8) with the dilute acid at a temperature below that of the room. This temperature must not, however, be so low as to arrest the chemical action. The process of solution may be greatly accelerated by the use of a platinum-stirrer;[1] but a brass stirrer coated with asphaltum may be employed (see ¶ 92, 1). The quantity of dilute acid used must be found by weighing the calorimeter with and without it; and the rise of temperature of this acid must be determined by a series of observations of temperature (¶ 92, 10) both before and after the experiment. It is well also to re-weigh the calorimeter after the experiment, to guard against any loss of material (¶ 92, 5). The loss of weight due to the escape of nitric oxide gas will hardly be detected.

(2) ZINC OXIDE AND NITRIC ACID. — The experiment is now to be repeated with a quantity of zinc oxide which would be formed by the combustion of 1 gram of zinc. This quantity is 1.25 $g.$, very nearly. The same weight and strength of acid are to be used as before (1); but the temperature should be very little below that of the room.

---

[1] Currents of electricity generated by the contact of platinum and zinc assist the chemical action. It is, indeed, stated by some authorities that in the absence of such currents *perfectly pure* zinc is not attacked by dilute acids.

The density of the acid used should be determined roughly as in ¶ 40.

From the results of this experiment the student is to calculate (as in ¶ 106, below) the number of units of heat given out by 1 gram of zinc in uniting with an excess of dilute nitric acid, also what part of this heat is due to its uniting with the oxygen of the acid. The heat of combination of zinc with nitric acid will be found to have an important bearing upon problems relating to electric batteries in which zinc is the dissolving element and nitric acid the oxidizing agent (§ 145).

¶ 106. **Calculations relating to Heat of Combination.** — It is necessary, in general, to find the specific heat of the liquid used for a determination of heats of combination (see Experiment 34). The specific heats of certain solutions, amongst them nitric acid, may be found, when their densities are known, by Table 30. In calculating the thermal capacity of a calorimeter, the specific heat of the glass composing the inner cup may be taken as 0.19.

If w_1 is the weight of zinc employed, s_1 its specific heat (.095), t_1 its original temperature; if w_2 is the weight of acid employed, s_2 its specific heat (from Table 30), and t_2 its original temperature reduced (see ¶ 93, 2) to the time of solution; if c is the thermal capacity of the calorimeter, t_3 its original temperature (the same as t_2) and t the temperature of the mixture, we have for the heat of combination h (substituting h for l in the general formula of ¶ 100, and changing signs, since h would be negative if heat were absorbed),

$$h = \frac{(w_2 s_2 + c)(t - t_2) + w_1 s_1 (t - t_1)}{w_1}. \quad \text{I.}$$

If, as in the experiment, a comparatively large quantity of acid is employed, the second term of the numerator may be neglected. When, moreover, 1 gram of zinc is used, $w_1 = 1$, and we have,

$$h = (w_2 s_2 + c)(t - t_2), \textit{ nearly.} \quad \text{II.}$$

The truth of the last formula is sufficiently evident, since s_2 is the thermal capacity of 1 gram of the acid, $w_2 s_2$ must be that of w_2 grams; and this added to the thermal capacity (c) of the calorimeter must represent (neglecting the 1 gram of zinc) the total thermal capacity. In the formula (II.) the total thermal capacity is simply multiplied by the number of degrees rise in temperature. This must give the number of units of heat developed by the combination of the zinc with the acid.

The heat of combination of zinc oxide may be calculated by formula I. To find the heat given out by a quantity of zinc oxide (1.25 grams, nearly) which contains 1 gram of metallic zinc, this heat of combination must be multiplied by 1.25. The same result may be obtained directly by formula II. if, as in the experiment described, we have employed 1.25 grams of zinc oxide.

The chemical reaction which takes place when zinc is dissolved in nitric acid may be divided theoretically into two stages: first, the combination of 1 gram of zinc with oxygen, which is obtained by the decomposition of a part of the nitric acid,[1] thus:

Zinc. Oxygen. Zinc oxide.
$$Zn + O = Zn\,O; \qquad (1)$$

and, second, the combination of the 1.25 grams of zinc oxide thus formed with more of the nitric acid to form zinc nitrate, thus:

Zinc oxide. Nitric acid. Zinc nitrate. Water
$$Zn\,O + N_2O_5 \cdot H_2O = ZnO \cdot N_2O_5 + H_2O. \quad (2)$$

We have already found the heat developed by the process as a whole. We have also found the heat developed in the second stage of the process, namely, the union of 1.25 grams of zinc oxide with nitric acid. The difference between these two quantities of heat must (by the principle of the conservation of energy) be equal to the heat developed by 1 gram of zinc in combining with oxygen extracted from nitric acid.

If, for example, 1 gram of zinc dissolving in 100 grams of nitric acid of a certain strength gives out

[1] Nitric acid, thus deprived of its oxygen, may be reduced to nitrous acid, nitric oxide (gas), or even to ammonic nitrate. The reactions are as follows:—

$2\,Zn + 3\,N_2O_5 \cdot H_2O = 2\,ZnO \cdot N_2O_5 + 2\,H_2O + N_2O_3 \cdot H_2O$ (nitrous acid).
$3\,Zn + 4\,N_2O_5 \cdot H_2O = 3\,ZnO \cdot N_2O_5 + 4\,H_2O + 2\,NO$ (nitric oxide).
$4\,Zn + 5\,N_2O_5 \cdot H_2O = 4\,ZnO \cdot N_2O_5 + 3\,H_2O + (H_4N)(NO_3)$ (ammonic nitrate).

Nitrous acid may be formed by the reduction of strong nitric acid. The presence of nitric oxide gas may usually be recognized by the red fumes which are generated when nitric acid is reduced. Ammonic nitrate is formed only in very weak solutions (Wurtz, Chimie Moderne, p. 169).

1,500 units of heat, while an equivalent (1.25 grams) of zinc oxide gives out only 400 units of heat, it is evident that 1500 − 400, or 1100, units of heat are due to the combination of 1 gram of zinc with the oxygen of the acid.

¶ 107. **Heat of Combustion.**— We have seen in the last section how we may find indirectly the amount of heat given out by a gram of a given material when it combines with the oxygen of an acid. This heat varies greatly according to the difficulty of extracting the oxygen in question. If, for instance, as in sulphuric acid, the oxygen must be taken away from hydrogen, for which it has a great affinity, nearly three fourths of the energy will be spent in decomposing the acid. In the case of nitric acid, less difficulty is encountered; since nitric acid is more readily decomposed (see footnote, ¶ 106). Even, however, in the case of chromic acid, in which the oxygen approaches very nearly its condition in the free state, the heat of combination with oxygen will differ somewhat from the result which we should obtain by burning a metal in oxygen gas.

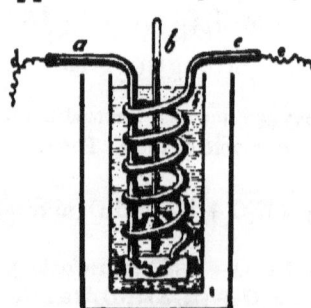

Fig. 84.

The heat given out by one gram of a substance when burned in oxygen is called its heat of combustion in oxygen. It may be determined directly by an apparatus shown in Fig. 84. The substance in question is placed in a deflagrating spoon, i, contained in a water-tight chamber, h; oxygen (or air) is admitted to this chamber by the tube a, and the gaseous products of combustion, if any, escape through the spiral tube gfc. The whole system of tubes is surrounded by water, contained in a calorimeter of the ordinary sort. When the temperature of the water has been observed, the substance is ignited by a current of electricity. From the rise of temperature and the thermal capacities of the calorimeter and its contents, the heat of combustion is calculated.

To determine the heat of combustion of a gas with this apparatus, a third tube must be added to supply the gas. A much simpler device consists, however, of a small metallic cone soldered into the bottom of a calorimeter. The cone ends above in a spiral tube, surrounded by water. A gas jet burned beneath this cone will give up nearly all of its heat to the water. The quantity of gas used is measured by a gas-meter. The determination of heats of combustion in general is an exceedingly difficult problem, but the ambitious student may be encouraged to attempt a rough determination of the heat of combustion of coal-gas or alcohol with a simple apparatus like the one described.

EXPERIMENT XXXIX.

RADIATION OF HEAT.

¶ 108. **The Pyroheliometer.** — A simple form of pyroheliometer ($\pi\hat{v}\rho$, fire, heat; $\H{\eta}\lambda\iota o\varsigma$, sun; $\mu\acute{\epsilon}\tau\rho o\nu$, measure), or instrument for measuring the heat radiated by the sun, consists of a hollow tin box (Fig. 85) filled with water. One of the outer surfaces of the box is blackened, so as to absorb most of the heat which falls upon it. This surface is turned perpendicularly to the rays, the intensity of which is to be measured. The temperature of the water is observed by a thermometer passing through a hole in the side of the box. The number of heat units absorbed is calculated from the rise of temperature and thermal capacity of the vessel and its contents, as in other experiments in calorimetry. An allowance for cooling is made by watching the thermometer when the instrument is in shadow. It is found in this way that the solar radiation may amount to nearly 2 units of heat per minute on each square centimetre of surface.

Fig. 85.

The pyroheliometer may also be used to measure the heat radiated by a candle, or any other source of heat; or it may be employed simply to compare two sources with each other. In all such experiments it is obvious that the distance of a given source of heat must be taken into account. It will be found, for instance, that the heat radiated by an ordinary candle-flame at a distance of about 2 $cm.$ may be as intense as the sun's heat. At the distance of a decimetre, the heat from the candle could hardly be detected by a pyroheliometer.

¶ 109. **Application of the Law of Inverse Squares.** When a person stands midway between two sources of heat which are equal in every respect, he feels of course equal intensities of radiation. If, however, one of these sources is much more powerful than the other, he must approach the smaller of the two in order that the warmth from both may seem to be the same. Let the power of the first source be x, and the distance from it a; let the power of the second source be y, and the distance from it b; then according to the law of inverse squares (§ 94) the effects of the two sources will be proportional to $x \div a^2$ and to $y \div b^2$, respectively. If the two effects are equal, it follows that

$$x \div a^2 = y \div b^2; \text{ or } x : y :: a^2 : b^2.$$

It thus appears that the powers of any two sources of radiant heat are to each other *directly as the squares* of the distances at which they produce equal effects.

The same reasoning may be applied to two sources of light, to two sources of sound, or to any two sources of radiant energy, the effect of which diminishes as the square of the distance increases.

We have, accordingly, a principle by which we may compare any two sources of energy of the same kind; namely to find two distances, a and b, at which equal effects are produced.

To test the equality of two effects with any degree of precision, it is necessary to employ a "differential" instrument of some sort; that is, an instrument which is constructed especially to indicate the difference between two effects. The instrument must be so delicate that in the absence of any indication, we may assume that the two effects are equal. The methods for the comparison of two sources of heat about to be described, will be found to belong to the general class known as "null methods" (§ 42).

¶ 110. **The Differential Thermometer and the Thermopile.** — I. A differential thermometer, useful for the comparison of two sources of radiant heat, may be constructed as follows: two cylindrical metallic boxes, d and e, about 10 $cm.$ in diameter, and 1 $cm.$ deep, are made out of the thinnest brass, and fastened by a layer of wax to the support bh. The glass U-tube or gauge, fg, contains a little colored liquid, and is attached by rubber couplings to the boxes d and e, so that the system may be air-tight. The outer faces of the boxes, d and e, are coated with lampblack, to absorb heat; the sides may be covered with wool to prevent loss of heat. The two conical shields, a and c, blackened inside, are finally added to cut off lateral radiation.

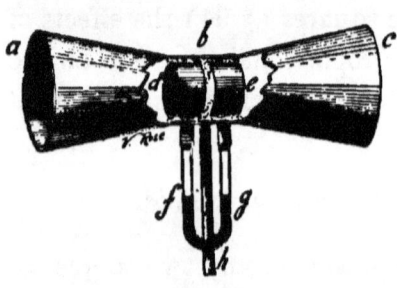

FIG. 86.

A very slight amount of heat falling on the blackened surface of either of the cylinders, d or e, will cause an expansion of air within the cylinder in question. Unless this is offset by an equal expansion of air due to an equal amount of heat falling on the other cylinder, the level of the liquid in the gauge fg will be affected.

II. An instrument which may be made much more sensitive than a differential thermometer is represented in Fig. 87, and in de, Fig. 88. It consists of an alternate series of strips of bismuth and antimony, joined together in a sort of zigzag. Only four strips are shown in the figure, but a much greater number is generally used. The combination is known as a "thermopile," or "heat-battery." It is usually mounted on a support (Fig. 88),

Fig. 87.

Fig. 88.

and provided with two conical shields, a and c. When heat falls on either set of junctions, as d, a current of electricity is generated (see Exp. 95). This current is measured by a galvanometer, f, the terminals of which are connected by wires with the terminals of the thermopile. The deflection of the galvanometer needle is reversed if heat falls on the opposite face of the thermopile, *e*. When equal amounts of heat fall on both the faces, *d* and *e*, the needle should not be deflected.

It would be out of place here to discuss the principles which underlie the phenomena in question. The student should for the present regard a thermopile and galvanometer simply as a convenient substitute for a differential thermometer and U-tube.

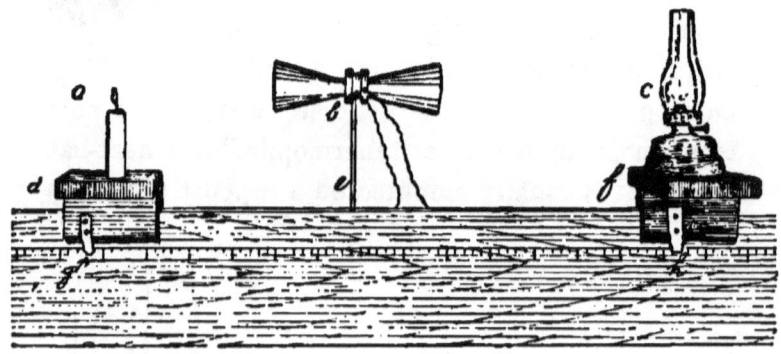

Fig. 89.

¶ 111. **Determination of Candle-Heat-Power.** — A thermopile connected with a galvanometer, as in Fig. 88, is mounted on a fixed support (*be*, Fig. 89), in the middle of a horizontal graduated rail (*gh*). The needle of the galvanometer is made to point to zero (¶ 112, 7). Two movable supports, *d* and *f*, constructed so as to slide along the rail, are placed one on each side of the thermopile. A candle (*a*) and a small kerosene lamp (*c*) are then mounted on the supports, d and f respectively, so that the flames may be on a level with the thermopile (¶ 112, 5). The supports are then to be set permanently at such distances from the thermopile (¶ 112, 2) that either flame alone will cause a deflection of the galvanometer of at least 45° (¶ 112, 1), but that both together will cause little or no deflection. The height of the lamp-flame is then adjusted, if necessary, until the deflection is reduced to zero.

The lamp and candle while still burning are next to be weighed as accurately as possible on a pair of open scales (Fig. 1, ¶ 2), and the time of weighing is to be noted in each case. The lamp and candle are then returned to their former positions on the supports d and f, where they are allowed to burn for, let us say, half an hour.

Meanwhile the distance of each from the nearer face of the thermopile is accurately determined by means of the markers (g and h), which should be just under the centres of the flames (¶ 112, 3). The distance (de, Fig. 88) between the faces of the thermopile must also be measured and allowed for (¶ 112, 4). If the needle of the galvanometer shows any deflection in the course of the experiment, it must be brought back to zero by increasing or diminishing the flame of the lamp. At the end of the half-hour, the candle and lamp are to be re-weighed in the same order as before, *while still burning*.

The candle and lamp are now to be replaced on their supports (d and f respectively), each of which is to be set permanently at the same distance from

the thermopile as before, but on the other side of it
(¶ 112, 8). The height of the lamp-flame is to be
adjusted so as to neutralize the heat from the candle;
and at the end of another half-hour, the lamp and
candle are to be re-weighed, as before, while still
burning.

Instead of a thermopile, a differential thermometer
(¶ 110) may be employed, with essentially the same
precautions (see ¶ 112). Instead of
the kerosene lamp, an electric incandescent
lamp may be used (Fig. 90). In
this case it is necessary that the zinc-
plates of the battery furnishing the
electricity for the lamp should be
weighed before and after the experiment.
These plates should be well
amalgamated with mercury to prevent
unnecessary loss of material.

Fig. 90.

In any case the candle-heat-power of
the lamp is to be calculated and reduced to the
standard rate of consumption, as will be explained
in ¶ 113.

¶ 112. **Precautions in the Determination of Candle-
Power.** — (1) Before attempting an accurate comparison
of two sources either of heat or of light, it is
well to make sure that the instrument to be employed
is sufficiently sensitive (§ 42). For this purpose it is
first exposed to the radiation from the feebler source
alone. To make a comparison, for instance, accurate
within 1 per cent, the response must be 100 times
as great as the minimum perceptible. The sensi-

tiveness of the combination should, if necessary, be increased by bringing the source in question closer to the instrument until a sufficient response is obtained.

(2) It is important that one of the two sources compared should be at a fixed distance from the instrument throughout an experiment. When an oil-lamp or gas-flame is one of the sources, so that the height of the flame can be adjusted, it is well that both sources should be fixed; and for convenience in calculation, each distance may be made equal to some round number.

(3) The distance of the sources from the instrument may be most conveniently determined by means of markers (g, h, in Fig. 89). These markers should be in line with the centre of the source of light or heat (as, for instance, h), not at one side of it (like g). The student should confirm the indications of the markers by direct measurements. It should be remembered that the distances sought lie between the centre of a flame and the surface illuminated by it.

(4) Care must be taken in measuring the distances ad and ec to allow for the distance de (Figs. 86 and 88) between the two surfaces illuminated. This distance should be determined by direct measurement; for this purpose the conical shields must of course be removed.

(5) It is important that the rays of light or of heat should be equally inclined with respect to the two surfaces d and e. To help in securing this result, the surfaces should be made vertical, and the

sources of light or heat should be raised or lowered until they are on a level with these surfaces. Neither angle of incidence should exceed 20°. In this case slight differences in the angles of incidence, as in Figs. 96 and 98, will have no perceptible effect on the result.

(6) The conical shields a and b (Figs. 86 and 88) will serve to cut off lateral radiation. It is, however, necessary to place large black screens *behind* two sources of light which are being compared, so as to shut out light from all other sources. A dark room is of great service in photometry; a room of uniform temperature is equally important in measurements of radiant heat.

(7) Before comparing two sources of heat or light, it is well to make sure that the instrument to be employed is not affected by radiation from the windows or from the walls of the room (§ 32). The liquid in a differential thermometer should stand at the same level, for instance, in both arms of the gauge. If it does not, the gauge should be temporarily disconnected so that the air-pressure may be equalized. The needle of a galvanometer connected with a thermopile should point to zero, otherwise it should be made to do so by twisting the thread by which it is suspended, or by placing a magnet in its neighborhood. If the two surfaces of a photometer do not appear equally dark, it is necessary to make a rearrangement of the screens, by which at least equality of illumination may be secured.

(8) To eliminate all errors arising from unequal

radiation from surrounding objects, and from any inequality in the surfaces illuminated, two determinations should always be made (see § 44). In one of these a given surface is illuminated by the weaker source of light or heat; in the other, it is illuminated by the stronger source. An error in the adjustment of the markers may also be eliminated in this way.

¶ 113. **Calculations relating to Candle-Power.** — The standard candle is defined as one, seven-eighths of an inch in diameter (six to the pound), burning 120 grains of spermaceti per hour. A paraffine candle does not give out quite so much light as a sperm candle under similar circumstances. It is thought that no perceptible error will be committed by substituting for a standard candle one of paraffine burning 8 grams per hour ($123\frac{1}{2}$ grains, nearly). An ordinary candle may of course burn a little more or less than the standard. Since the heat or the light is very nearly proportional to the rate of consumption, we find that the actual candle-power of a paraffine candle[1] is equal to one eighth the weight in grams of the paraffine burned in one hour. This gives us the quantity, x, in the formula of ¶ 109. Hence, if a lamp at a distance b has the same effect as x standard candles at the distance a, as regards either heat or light, we may find the number of standard candles, y, to which this lamp is equivalent by the formula —

$$y = \frac{b^2}{a^2} x.$$

---

[1] The heat radiated in all directions by an ordinary candle amounts to about 2 units per second. This is only a small part of the total

By the "candle-power" of a lamp is ordinarily meant the number of standard candles to which it is equivalent in respect to light (see Exp. 40). The number of candles to which it is equivalent in respect to the radiation of heat may be called its "*candle-heat-power*." It is evident that the thermopile and the differential thermometer, which absorb all rays alike (whether visible or invisible), are instruments for determining the candle-heat-power as distinguished from the candle-light-power of any source.

It is interesting to reduce the candle-power of a lamp to the normal rate of consumption of a candle (8 grams per hour). We first divide the actual candle-power of the lamp by the number of grams burned in one hour to find the candle-power corresponding to 1 gram per hour; then we multiply the result by 8. A surprising similarity exists between the candle-powers of different materials when thus reduced to a common standard.

EXPERIMENT XL.

PHOTOMETRY.

¶ 114. **Determination of Candle-Power by means of a Photometer.** — I. BUNSEN'S PHOTOMETER. — A very fair comparison of two sources of light may be made by means of a scrap of white paper rendered trans-

quantity of heat generated by combustion, which amounts to about 20 units per second. Less than 4% of the radiant heat is visible as light.

lucent at the centre by a drop of oil or varnish. When such a scrap is held up in front of a light, the oil-spot appears bright, as in Fig. 91; when held behind a light, it looks dark, as in Fig. 92. If both

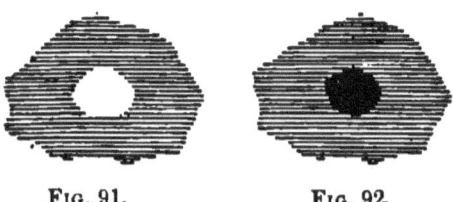

Fig. 91. Fig. 92.

sides of the paper are equally illuminated, the spot may nearly or quite disappear. Usually, however, the oil-spot seems a little darker than the rest of the paper. It is necessary, therefore, to look at it from both sides. When it appears equally dark from both points of view, we may infer that the two sides of the paper are equally illuminated.

To make use of an oil-spot for a comparison of two lights, the paper (*b*, Fig. 93) is provided with

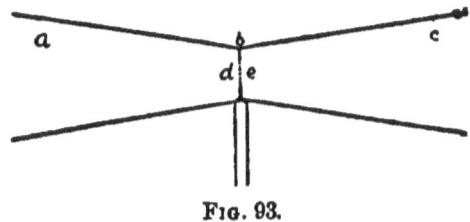

Fig. 93.

two shields, *a* and *c*, to cut off lateral radiation, and is mounted in the place of the thermopile (*b*, Fig. 89, ¶ 111) between a candle, *a*, and a lamp, *c*. The lamp-flame is adjusted as in ¶ 111 until the paper seems equally illuminated on both sides, *d* and *e*.

The distances of the lamp and candle, and the weights burned in one hour by each are found in the same manner as with the thermopile.

In practice the form given to the shields is not generally conical, as in the case of a thermopile, but barrel-shaped (see Fig. 94). The object of this is to facilitate the examination of the oil-spot through two openings, *a* and *c*. Such an instrument is called a Bunsen's photometer.

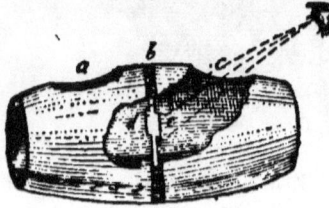

Fig. 94.

The general precautions in the use of a photometer have already been enumerated (¶ 112). Certain special precautions will be considered in ¶ 115. The results are to be reduced as in ¶ 113.

II. RUMFORD'S PHOTOMETER. — If the diaphragm and shields used in Bunsen's photometer (Fig. 93) are removed, leaving only the rod by which they were supported, and if a piece of paper (*ac*, Fig. 95) is fastened to this rod so as to be equally inclined to the rays falling upon it from the lamp and from the candle (Fig. 89); then when the flames are placed at such distances as to give equal amounts of light at the point *b*, the shadows *ab* and *bc* (Fig. 95) cast by the rod should be equally dark. The instrument, thus arranged, is a form of Rumford's photometer, depending upon the principle that equal illuminations cause equal shad-

Fig. 95.

ows; it might be substituted for a Bunsen's photometer for a rough comparison of two lights.

It is obvious, however, that a slight inclination of the paper might expose it very unequally to the rays from the two sources, and thus vitiate the results. To lessen errors from this source, both lights are in practice placed on the same side of the rod, b (Fig. 96), the two shadows of which, d and e, are thrown horizontally on the vertical surface, de. When these shadows have been made equally dark by adjusting the distances of the lamp and candle, or the height of the lamp-flame the two lights are to each

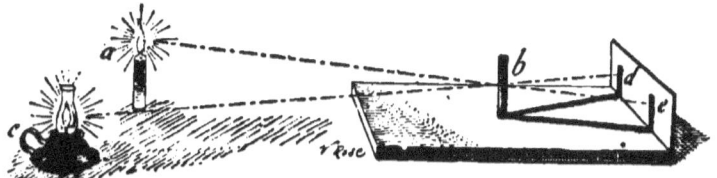

FIG. 96.

other as the squares of the distances ae and cd. These distances are therefore to be measured.

The student should observe that the distance of the rod from the screen may affect the sharpness of the shadows, but not their darkness, which depends simply on the distance of the lights *from the screen*. It is well to have the rod close to the screen, in order that the two shadows may be near together, but not so close that the shadows overlap. A small amount of light from the windows need not vitiate the result, provided that it casts no shadow on the screen. If it does, the light must be cut off.

The weights burned by the lamp and candle in one hour are found as with a Bunsen's photometer (I.), or with a thermopile (¶ 111); and the results are reduced in the same manner (¶ 113).

III. Box Photometer. — Instead of using a rod, as in Rumford's photometer (Fig. 96), it is sometimes

Fig 97.

advantageous to employ a partition (b, Fig. 97). One half (d) of the screen, de, may thus be illuminated by the candle (a), and the other half (e) by the lamp (b). The screen is made translucent, so that the intensities of illumination may be compared with the eye behind it.

This form of photometer is particularly useful when it is possible to enclose the whole apparatus in a box. A horizontal section of such a box is shown

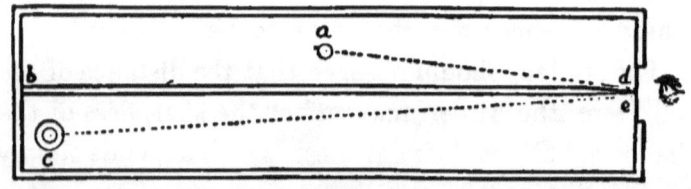

Fig. 98.

in Fig. 98. The distances ad and ce are measured directly by a metre rod.

If the angles of incidence, adb and ceb, differ by more than 10°, it may be well to alter the screen de slightly so that its inclination to both rays may be the same (¶ 112, 5).

An arrangement in which the distances of the lamp and candle are adjustable is represented in Fig. 99, which gives a view of the apparatus from above. The lights are contained each in one of the sliding boxes, e and f. The top of the main box is closed as far as the ends of the sliding boxes by a set of covers (b, c, and d). All direct light is thus excluded from the photometer. A cloth cover, a, may be thrown over the head when it is desired to compare very feeble illuminations.

Box photometers may also be constructed on Bunsen's or on Rumford's principle. They have the advantage of a dark room without its expense or inconvenience.

The determinations of candle-power, and the reduction of the results, are made in precisely the same manner as in II. with a Rumford's photometer. See also ¶ 113.

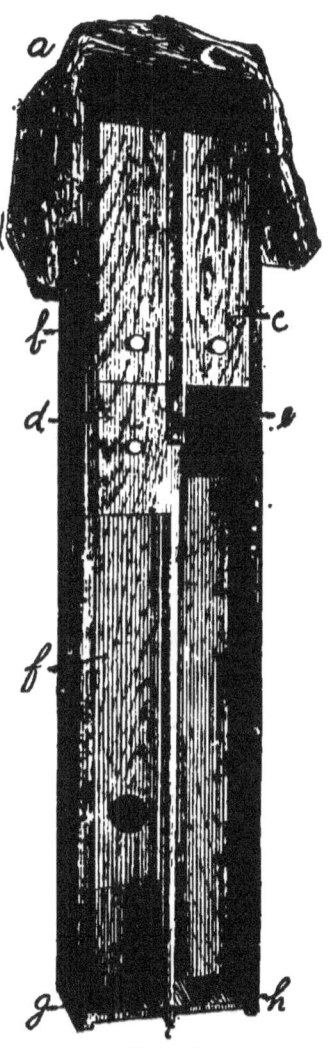

FIG. 99.

¶ **115. Errors in Photometry due to Color Blindness.** — Light is essentially a physiological as distinguished from a physical quantity. There is no standard by which we may prove that one kind of light is more brilliant than another. A person who is "color-blind" may consider a blue light brighter than a red light, which to a person of "normal vision" may seem much the brighter of the two. All eyes are in a certain sense color-blind, since the greater part of the rays which fall upon them are wholly invisible.

The modern theory of color may be stated briefly as follows: There are three principal effects produced on the eye by rays of light. The first is to excite in the retina a sensation which we call red. This is due mostly to waves of light between 60 and 70 millionths of a centimetre in length. The second is to excite a sensation which we call green. Nearly all rays of light produce this effect (green) to a certain extent; but it is caused most strongly by waves between 50 and 60 millionths long. The third effect is a sensation which we call violet, due to waves from 40 to 50 millionths in length. When waves[1] of different lengths are mixed, complex sensations are produced. Red and green rays together may produce, for instance, a sensation which we call yellow; violet and green may produce blue; red and violet may produce purple; while red, green, and

---

[1] The student must distinguish carefully the effects of mixing waves of light from the effects of mixing paints. These effects are in a certain sense opposite.

violet rays together may cause the sensation which we are familiar with in ordinary white light. Again, a single wave may produce two sensations: one 60 millionths of a centimetre long will, for instance, produce the double sensation which we call yellow; while one 50 millionths long will appear blue. The various hues which we find in different objects are due to the proportions, simply, in which the sensations of red, green, and violet are excited. The eye is capable of no fourth sensation by which the effect can be modified. According to this theory, two lights should be compared, (1) by means of the red rays, (2) by means of the green rays, and (3) by means of the violet rays which they emit.

The simplest way to compare the candle-power of two lights with respect to red rays is to hold a piece of ordinary " ruby glass " before the eye in observing the brilliancy of the two surfaces illuminated. Green and violet glasses may similarly be employed for the green and violet rays; but pure violet glass can hardly be obtained. It is better to use a piece of ordinary glass stained with violet-aniline containing a trace of Prussian blue.

With these precautions, personal errors in photometry might undoubtedly be diminished, particularly in the comparison of lights of different hues or tints; but as long as the eye alone is used to compare the brilliancy of two surfaces, it is doubtful whether the errors of a photometric comparison can ever be greatly reduced. The " probable error " of such a comparison may be estimated at about 5 per cent.

EXPERIMENT XLI.

PRINCIPAL FOCI.

¶ **116. Determination of the Principal Focal Length of a Converging Lens.** — The principal focal length of a lens may be defined (see § 103) as the distance at which it brings parallel rays to a focus. An "optical bench," convenient for the measurement of focal lengths, is represented in Fig. 100. It consists of a wooden plank, set up edgewise, with three sliding supports, d, e, and f, the positions of which are deter-

Fig. 100.

mined respectively by the markers g, h, and i. The apparatus is in fact the same as, or similar to, one already employed in Experiments 39 and 40 (see Fig. 89).

(1) ORDINARY METHOD. — To find the principal focal length of a lens, it is mounted (see b, Fig. 100) on one of the slides (e), directly over the marker (h) (see ¶ 112, 3); and a translucent screen (c) is attached to another slide (f) directly over the marker (i). The third slide (d) is temporarily removed, so that the rays from distant points (at the

left of the figure) may be focussed by the lens (*b*) on the screen (*c*). That this may be possible, the bench should be set up in front of an open window commanding a distant view.[1] Either houses or trees may afford suitable images. It is assumed, however, that the objects in question are so far off that rays from any point in these objects may be considered parallel. They should be at least a hundred times as far from the lens as the lens is from the screen.

The distance between the lens and the screen is to be adjusted so that the image thrown on the screen may be as distinct as possible. The image may be viewed either from in front or (since the screen is translucent) from behind. The number of details visible in the image is the test of its distinctness most easily applied. When difficulty is found in the precise adjustment of distance, the screen is first brought so near the lens that the most minute details disappear; then it is placed so far from the lens that the same result is obtained. Midway between these two positions is the principal focus of the lens.

The distance of the principal focus from the centre of the lens is taken as the measure of its principal focal length. It is determined by observing the positions of the two markers, *h* and *i*, with respect to the scale close behind them. If either of the markers is out of line with the lens or screen, as the case may be, an error will evidently be introduced into the result

---

[1] In the absence of any suitable object, we may use a projecting lantern, focussed so as to give parallel rays. To obtain this result, the slide must be placed in the principal focus of the projecting lens.

(¶ 112, 3). To eliminate this error, we may interchange the places of the lens and screen. The whole bench must then be turned round so that an image may be formed by the lens on the back of the screen. The thickness of the screen should be so small that it need not be taken into account. If either of the markers is out of line, the distance between the lens and screen will apparently be increased in one case but diminished in the other case, and by an equal amount. The average of the two distances indicated by the markers is, therefore, the true distance from the centre of the lens to the screen.

If there is a second scale on the farther side of the bench, there will be no need of turning it round. We have only to turn round the slides e and f.

It is well to confirm the accuracy of the scale or scales in question by a direct measurement between the thin edge of the lens and the screen. The measuring rod must be held perpendicular to the screen, as in Fig. 100. One measurement should be taken from the farther edge of the lens, another from the nearer edge, and a third from the top of the lens. If any marked differences are observed, the lens should be readjusted until these differences disappear.

(2) METHOD OF PARALLAX. — Instead of using a screen (c, Fig. 100), we may employ a wire netting or simply a vertical wire. If the wire coincides in position with the image formed by the lens, no "parallax" (§ 25) will be apparent when the eye is moved from side to side. If the wire is behind the image, it will seem to follow the eye; or if it is in

front of it, it will always appear to move in the opposite direction (see diagrams, Fig. 103, ¶ 118). The phenomena of parallax afford in fact a very delicate test by which a wire may be placed exactly in the image, and the position of the image thus accurately determined. This is called focussing by the method of parallax. The distance of the image from the lens is found from the indications of the markers, and confirmed by direct measurements as before (see 1).

(3) INDIRECT METHOD. — Another way of finding the principal focus of a lens involves the use of a telescope, which has been adjusted so that parallel rays striking the object-glass (g, Fig. 101) are brought to a focus at a point c where cross-hairs are placed.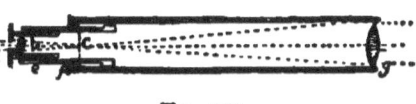

FIG. 101.

The first step in focussing a telescope is always to make the distance of the eye-piece (b) from the cross-hairs (c) such that the latter may be seen as clearly as possible through the opening a. This is done by sliding the tube d within the tube e. Then the tube e is pushed into or drawn out from the tube f so that the cross-hairs may coincide with the image at c. In the last adjustment, care must be taken not to disturb the distance of the eye-piece from the cross-hairs, unless, as sometimes happens, the focus of the eye has changed so that the cross-hairs are no longer visible; in this case the first adjustment must be repeated before the second can be made. In some telescopes the method of focussing by parallax (see 2) can be used, but generally we have to depend simply on the distinctness of the image (see 1). If the telescope is accurately focussed, the image and the cross-hairs should both appear distinct to the eye.

A telescope thus focussed is mounted as in Fig. 100 at any point, a, in front of a lens, b. It will probably be found that a page of fine print replacing the screen, c, may be easily read through the combination. The distance of the page from the lens should be varied if necessary, so that the print may seem as distinct as possible.

The student should note that, owing to the parallelism of the rays from a given point in passing between the lens and the telescope, the distance between the lens and telescope does not affect the focus.

The principal focal length of a lens has been defined as the distance from the lens at which parallel rays are brought to a focus; it might also have been defined as the distance from an object at which rays diverging from it are rendered parallel by the lens. It is evident that the rays diverging from any point of the printed page (c) must be rendered parallel by the lens (b) in order to be visible in the telescope (a); for this telescope has been focussed for parallel rays, and cannot, therefore, be in focus for any others. It follows that the distance from the lens to the screen is equal to the principal focal length of the lens; the latter is, therefore, to be measured as in the methods previously described (see 1 and 2).

(4) COLOR METHOD. — Instead of depending entirely upon the distinctness with which the print can

be read, we may observe the colors with which each black letter seems to be surrounded. Unless the lens is of peculiar construction, so as to focus all rays alike, it will be found impossible to avoid this phenomenon. Let us suppose that the red rays are accurately focussed; then the green and violet rays will be just out of focus, and hence somewhat scattered. The spaces which would otherwise be perfectly black will, therefore, have a bluish tinge (¶ 115), particularly near the edges of the letters. In the same way, if the violet rays are just in focus, reddish or yellowish borders will encroach upon the spaces in question. It is thus evident that the principal focus of a lens depends upon the kind of light employed. Green light may be taken as the standard. To focus for the green rays, the distance of the lens from the print must be such that the black spaces have very narrow borders of a neutral tint; that is, one which inclines neither to red nor to blue.

To obtain the best results with the color-method, a perforated metallic lamp chimney should be substituted for the page of print (see Exp. 42). The measurements of distance are made and reduced as in methods previously described (see 1, 2, and 3).

The student should make at least two determinations of the principal focal length of a lens, — one by the ordinary method, the other by the indirect method, (3). The other methods will be met in experiments later on. The results of different methods should agree within limits which may be attributed to errors of observation.[1]

EXPERIMENT XLII.

CONJUGATE FOCI.

¶ 117. **Determination of Conjugate Focal Lengths of Lenses.** — A screen, c (Fig. 102), and a lens, b, are to be mounted on movable supports, as in Exp. 41; but in place of the telescope (a, Fig. 100) the support, d, is to carry a lamp, a, having a metallic chimney with several small holes in it. The marker, g,

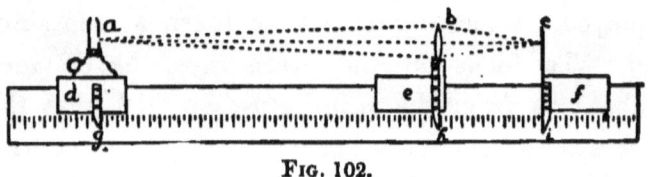

Fig. 102.

must be in line with the perforations in the chimney, not, as in ¶ 112, (3) with the flame, since the former and not the latter will be focussed upon the screen.

---

[1] If in (1) or (2) the object is too near, so that the rays from it striking the lens are perceptibly diverging, the distance of the screen from the lens must evidently be increased in order that these rays may be focussed upon it. On the other hand, if in 3 or 4 the telescope is focussed upon *the same object*, the distance of the print from the lens must be diminished in order that the rays which pass through the lens may be slightly divergent; for the telescope, being focussed for slightly divergent rays, can be in focus for no others. By averaging a result obtained by (1) or (2), with a result from (3) or (4), the true value of the principal focal length may be calculated, even when a distant view cannot be obtained.

Throughout this experiment the color method of focussing (see ¶ 116, 4) is to be used.

(1) The lens is first placed in the middle of the bench gi, with the lamp at a distance from it equal to twice its principal focal length, determined in Experiment 41. The screen is then moved until an image of the perforations of the chimney appears upon it; the distance between the lamp and screen is then measured. The lens will probably be found to be about half-way between the lamp and screen; if it is not exactly in the middle, it should be placed there, and the focus, if necessary, readjusted by increasing or diminishing the distance of both the lamp and the screen by an equal amount in each case. The distance of the screen from the lamp will be about four times the principal focal length of the lens.

(2) The lamp and screen are next separated by a distance equal to about five times the principal focal length of the lens; and the lens is placed so that the chimney may be focussed upon the screen as before. Two positions will be found, — one nearer the lamp, the other nearer the screen (see Fig. 102). In the first position, the image of the chimney will be magnified; in the second it will be diminished in size (see § 104). The second image will be the more distinct; the first, unless carefully searched for, may even escape detection. The distances ab and bc are to be determined in each case.

(3) The lamp and screen are finally separated as far as possible; and, as before, the lens is placed so as to throw first a magnified and second a reduced

image of the chimney upon the screen. In both cases, the distances *ab* and *bc* are to be determined.

The distances *ab* and *bc* in each of the cases (1), (2), and (3), are called conjugate focal lengths (§ 103). They may be determined by the readings of the markers *g*, *h*, and *i*. In (1) the *sum* of the distances *ab* and *bc* is alone needed, and should be confirmed by a direct measurement with a metre rod. If the markers are found to be tolerably accurate, the readings of the scale in (2) and (3) need not be confirmed by direct measurement.

From the results of each adjustment, the principal focal length of the lens is to be calculated by the formula derived from that in § 103: —

$$f = \frac{ab \times bc}{ab + bc}.$$

The results should agree with those obtained in Experiment 41 within a limit which may be attributed to the *thickness of the lens*, which has been disregarded in the formulæ.

The student should notice that it is impossible to focus the lamp upon the screen (1) when the distance *ac* is less than four times the principal focal length of the lens, no matter where the lens is placed; (2) when the distance (*ab*) between the lamp and the lens is less than its principal focal length, no matter where the screen is placed; and (3) when the distance (*bc*) between the screen and the lens is less than its principal focal length, no matter where the lamp is placed.

It should also be noticed that in (2) and in (3) the distances *ab* and *bc*, at which a magnified image is produced, are equal respectively to the distances *bc* and *ab*, at which we obtain an image reduced in size; and that in every case the distance between two perforations in the chimney is to the distance between their respective images as the distance of the lamp from the lens is to that of the screen from the lens (§ 104).[1] It is hardly necessary to call attention to the fact that all the images are inverted.

EXPERIMENT XLIII.

VIRTUAL FOCI.

¶ 118. **Real and Virtual Foci of Mirrors.** — Rays of light may be brought to a focus by a concave mirror as by a converging lens. If in Fig. 102 (¶ 117) we substitute for the lens, *b*, a mirror with its concave surface turned towards the lamp, *a*, and at a sufficient distance from it, an inverted image of the lamp will be formed at a point *c*, between *a* and *b*. This image, which will be reduced in size, may be received upon a screen, provided that the latter is not so large as to cut off all light from the mirror. Again, if the screen (*c*) is at a sufficient distance from the mirror (*b*), a magnified image of the lamp may be thrown upon it by placing the lamp at some point,

---

[1] It is instructive to prove this by actual measurement. See Experiment 38 in the Elementary Physical Experiments, published by Harvard University.

a, between b and c (as in Fig. 104), provided that the lamp does not intercept all the rays reflected by the mirror towards the screen. In any case the *real image* (c) formed by the mirror is on the *same side* of the mirror (b) as the object (a), not as in the case of a lens, on the opposite side of it.

The distances ab and bc are called, as in the case of a lens, conjugate focal lengths. The principal focal length of a concave mirror may be found by determining the distance at which parallel rays (or rays from a sufficiently distant object) are brought to a focus, or by the formula of ¶ 117, applicable to conjugate focal lengths. These methods are particularly valuable in the case of mirrors whose curvature cannot be determined by means of a spherometer (Experiment 21). Evidently the focal lengths of a mirror depend solely on its curvature. The material of which it is composed does not, as in the case of a lens, have to be considered.

The images thrown by a concave mirror upon a screen are instances of real images. The image of an object seen in a plane mirror is a typical case of a virtual image (§ 104). If the eye is placed *behind* the mirror (where the image seems to be) no light whatever is perceived. A thermopile would feel no heat there, nor would photographic paper be affected. And yet, as far as points *in front* of the mirror are concerned, the optical, thermal, and photographic effects are the same as if a real object existed behind the glass.

The simplest way to locate a virtual image is by

VIRTUAL FOCI.

the method of parallax (¶ 116, 2). A short wire is mounted in place of the lamp (*a*, Fig. 102) on a support, *d*; a longer wire, *c*, is attached to the support *f*, and a piece of looking-glass is placed between the wires on a support, *e*, instead of the lens *b*. The height of the wires should be such that the point of the long wire, *c*, may be visible above the image of the wire *a*, reflected (as in Fig. 103) by the mirror. As the eye is moved from the farthest left-hand point (see 1 and 3, Fig. 103) at which both wires are visible, to the farthest right-hand point (see 2 and 4, Fig. 103), both *a* and *c* (one being really, the other virtually, behind the mirror) will move from

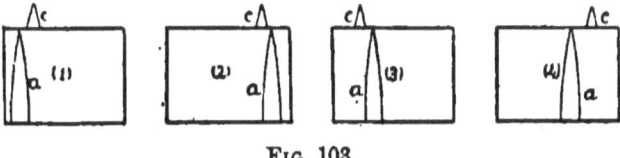

FIG. 103.

the left of the mirror to the right; but the one which is farthest off will apparently move farther than the other (see ¶ 116, 2). Thus if, as in (1) and (2), the point *a* moves completely across the mirror, while the point *c* only moves part way across it, we conclude that *a* is too far from (or *c* too near) the mirror, but if, as in (3) and (4), *c* moves wholly across while *a* moves only part way across, we conclude that *c* is too far from (or *a* too near) the mirror. By adjusting the distances *ab* and *bc* until no parallax (§ 25) is visible between *a* and *c*, the distance of the virtual image from the mirror may be determined.

It is found that the virtual image formed by a plane mirror is just as far behind it as the real object is in front of it.[1] If a mirror is slightly convex or concave, this will no longer be true. A comparison of the two distances ab and bc will serve therefore to detect any curvature in the surface of the mirror.

We notice that virtual images are never, like real images, inverted. When formed by a mirror they are always behind it. On the other hand, we shall see that the virtual focus of a lens is always on the same side as the object.

¶ 119. **Determination of Virtual Focal Lengths of Lenses.** — I. CONVERGING LENSES. — When the principal focal length of a lens exceeds the limit of the

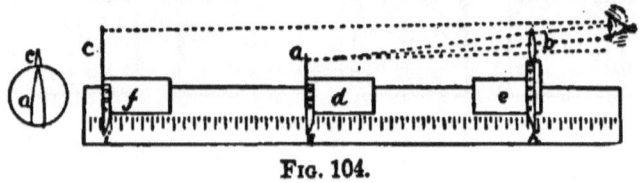

FIG. 104.

apparatus employed, it can be determined only by means of virtual foci. Two wires, a and c, are mounted on sliding supports, as in Fig. 104, on the same side of the converging lens (b) so that the top of the farther wire (c) may be visible just above the magnified image of the nearer wire (a) seen through the lens. The wires are then placed so that there may be no parallax (§ 25) between them when the eye is moved from side to side (see ¶ 118, Fig. 103). The virtual image of a then coincides with the real

[1] This may be shown by a simple geometrical construction. See Ganot's Physics, § 513, Deschanel, § 699.

point, c. The distances ab and ac are then measured, as in ¶ 117, and the principal focal length of the lens is calculated by the formula (see § 104),

$$f = \frac{ab \times bc}{bc - ab}.$$

II. DIVERGING LENSES. — With diverging lenses, focal lengths can be determined only by the method of virtual foci, since such lenses form no real images (§ 104). The method is essentially the same as that employed with converging lenses (see I.), except that the wire, a, viewed through the lens, b, must be further off than the wire, c, which is seen above or below it. It is well to substitute a broad netting or page of print for a, so that it may not be completely hidden by c.

The distances, ab and bc, are to be adjusted so that all parallax disappears between a and c; the virtual image of a will then coincide with c. The distances ab and bc are to be measured, and the value of f (which will be negative) is to be calculated by the same formula as before. It may be noted that a virtual image of distant objects is formed *between* a diverging lens and the objects in question, and at a distance (f) from the lens, which is sometimes called its (virtual) principal focal length.

The student should observe that a converging lens forms a virtual image *farther off* than the object looked at, while a diverging lens forms a virtual image *nearer* than the real object. Upon this fact depends in part the magnifying power of a converging lens, and the reducing power of a diverging lens.

The farther off an object is, the larger must it be in order that its image may occupy a given space on the retina; hence, the farther off we think it is, the greater will be our estimate of its dimensions.

In the arts, lenses are often numbered according to their principal focal length. A No. 12 spectacle lens is generally one which focusses distant objects at a distance of 12 inches. Near-sighted or diverging lenses are numbered on the same system. A No. 12 near-sighted lens combined with a No. 12 magnifying lens should form a perfectly neutral combination.

EXPERIMENT XLIV.

THE SEXTANT.

¶ 120. **Principle of the Sextant.** — A sextant may be constructed, as in Fig. 105, of two pieces of looking-glass, ag and aj, hinged together at a with their reflecting surfaces inward. The silvering is removed near e and near i, so that an object in the direction x may be seen through the two glasses; but enough silvering is left between b and j to make it possible also to see objects in the direction y, reflected by the mirror ag in the direction hi, then by aj in the

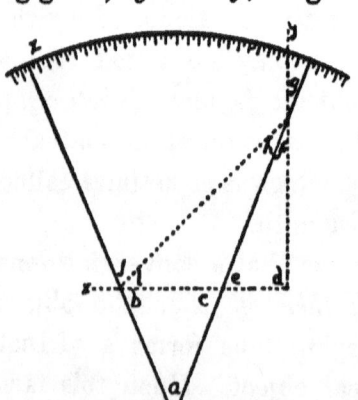

Fig. 105.

direction ie. The angle, a, between the mirrors may be measured by a graduated arc, yz.

Let us first find the relation between the angle d through which the ray y is bent and the angle a between the mirrors. The law of the reflection of light (§ 97) gives us the angles $b = j$ and $g = h$. The vertical angles c and e are equal by construction, also g and f; hence $f = h$. We have furthermore in the triangles abc and abh, —

$$a = 180° - b - c, \qquad (1)$$
$$a = 180° - b - i - h. \qquad (2)$$

Substituting equals for equals, we have, —

$$a = 180° - j - e, \qquad (3)$$
$$a = 180° - b - i - f. \qquad (4)$$

Adding (3) and (4),

$$2a = 360° - e - f - b - i - j;$$

or since b, i, and j together equal 180°,

$$2a = 180° - e - f. \qquad (5)$$

But from the triangle def, we have, —

$$d = 180° - e - f; \qquad (6)$$

hence, comparing (5) and (6), we find, —

$$d = 2a. \qquad (7)$$

We see, therefore, that when a ray of light is reflected by two mirrors, the angle (d) between its original direction (yd) and its final direction (xd) is equal to twice the angle (a) between the mirrors.

Now let us suppose that the plane *ayz* is made vertical, and that the angle *a* is adjusted so that the rays of the sun [1] from the direction *y* may seem, after being twice reflected, to come from the direction *x*, let us say that of the horizon; then the altitude of the sun is evidently 2*a*. The student should note that two objects in different directions may be visible *simultaneously* through a sextant. The sun may be made to appear, in fact, as if it were actually on the horizon.

¶ 121. Description of an Ordinary Sextant. — We have seen how a sextant may be constructed out

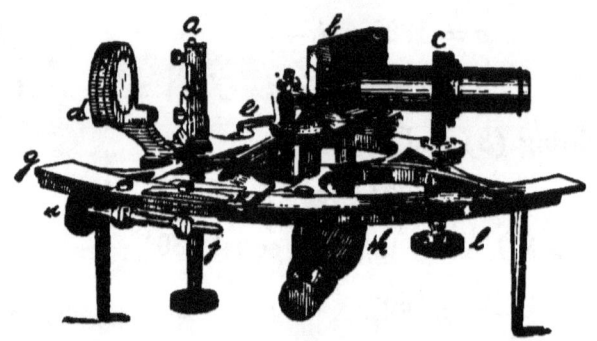

Fig. 106.

of two mirrors hinged together as in Fig. 105. In practice it would be necessary to remove most of the silvering between *j* and *z*, since it would otherwise interfere with the ray *yd* when the angle *d* is very small. In an ordinary sextant, this

---

[1] The mirror should be smoked near *f* and *g* before trying this experiment, in order that the brightness of the sun may be sufficiently diminished.

portion of the mirror is entirely removed. Of the two mirrors, az and ag, there remain in fact only the small portions, bj and hg, represented respectively by a and b in Fig. 106, or by ac and df in Fig. 107. The mirror a (Fig. 106) is fixed in position, and b is pivoted at its centre instead of an axis (as in Fig. 105) where the planes of the two mirrors intersect.[1] The angle between these planes is moreover measured, not by an arc (zy, Fig. 105) included by the angle (a), but by an arc g (Fig. 106) situated in quite a different part of the instrument. On this arc a vernier (h) connected by a movable arm with the mirror (b) serves to indicate the angles through which the mirror (b) is turned.

A tube or telescope, c (Fig. 106), permanently pointed toward the fixed mirror (a) serves principally as a guide for the eye. There is also, in most sextants, a set of dark glasses, d, which may be so placed as to diminish the light of the sun when looked at directly through the unsilvered part of the fixed mirror, a; there is also a set of dark glasses at e (not shown in the figure) to cut off excessive light reflected by the revolving mirror, b. A magnifying glass (f) is used for reading the vernier (h). The vernier is clamped by a thumb-screw (j), and slow motion is produced (only when clamped) by the tangent screw (i). There is also a screw (l) by which

---

[1] The fixed mirror, a, is called the "horizon-glass," because in nautical observations the horizon is usually seen through it; the revolving mirror, b, is called the "index-glass" because it carries the index. See Glazebrook and Shaw's Practical Physics, § 48.

the tube or telescope (*c*) may be either raised so as to come opposite the upper portion of the mirror, *a*, which is unsilvered, or lowered so as to be opposite the silvered portion. By this means, the relative brightness of the direct and doubly reflected images may be varied at pleasure. The handle *k* is of use especially in nautical observations.

¶ 122. **Adjustments and Reading of a Sextant.** — In order that a sextant may give accurate readings, certain conditions must be fulfilled.

(1) The tube or telescope, *c*, must be parallel to the plane of the graduated arc; for in demonstrating the relation between the angle (xdy, Fig. 105) through which a ray of light is bent and the angle (*a*) between the mirrors, we have assumed that the whole figure lies in one plane. This condition is fulfilled if a distant object, visible through the tube or telescope (*c*) in the middle of the field, appears, when sighted, to be in the same plane as the graduated arc. If this condition is not fulfilled, the position of the tube or telescope must be altered by an instrument-maker, so that the line of sight may be parallel to the plane of the graduated arc.

(2) The pivot on which the mirror (*b*, Fig. 106) rotates must be perpendicular to the plane of the graduated arc. This condition is fulfilled if the movable arm can be turned from one end of the arc to the other without either leaving it or binding against it. If it is not fulfilled, the sextant should be discarded.

(3) The revolving mirror should be perpendicular

to the plane of the graduated arc. This condition is fulfilled if the reflection of the arc in the mirror seems to be a continuation of this arc. If the reflected portion seems to slope upward or downward, the mirror leans forward or backward. The adjustment of the revolving mirror should not be attempted by the student, but should be left to the instrument-maker.

(4) The fixed mirror should be perpendicular to the plane of the graduated arc. This condition is fulfilled if, after the revolving mirror has been properly adjusted, the sextant can be set so as to give a single image of distant objects; for the fixed mirror is then parallel to the revolving mirror, and hence perpendicular to the arc. The reading of the sextant when so set is called its zero-reading (see ¶ 123). If no such setting can be made, the fixed mirror should be tipped a little forward or backward by turning one of the screws which hold it in place. This adjustment should be attempted only by persons who have acquired some skill in the use of a sextant.

(5) The fixed mirror should be nearly parallel to the revolving mirror when the index attached to the latter points to the zero of the graduated arc. This is the case if the sextant gives only a single image of distant objects when set as stated. If a double image is seen, one of the two mirrors should be rotated without disturbing the setting. A screw is usually provided for rotating the fixed mirror through a small angle. There is danger in so doing that the

last adjustment (1) may be disturbed. If it is, it must be repeated. The student is advised to omit the 5th adjustment altogether, since a slight error in it may cause a little inconvenience in allowing for zero error, but will not affect the accuracy of results.

(6) The arc and vernier must each be uniformly graduated. The uniformity of the arc may be tested (as in ¶ 48 *d*) by means of the vernier. If the latter subtends, for instance, 119 divisions in all parts of the arc, these divisions must have the same length. If the coincidences on the vernier follow in regular succession as the tangent screw (*i*) is slowly revolved, we may infer uniformity both in the main scale and in the vernier.

(7) The value of the main-scale and vernier divisions must be known. An accurate method of correcting the main scale will be considered (incidentally) in Experiment 45. To decide whether the divisions, of which every tenth one is usually numbered, are intended to be degrees, or only half-degrees, so as to represent the number of degrees through which a ray of light is bent (see ¶ 120, formula 7), a rough test will be sufficient. Thus if a string reaching from the pivot to the graduated arc also reaches from 0 to 120 on the arc, we may infer that the divisions are half-degrees. By calling them degrees we shall avoid the labor of doubling each reading of the sextant when measuring the angle through which a ray is bent by reflection in the two mirrors.

The divisions which represent degrees are divided in different instruments into two, three, four, six, and even twelve parts. The number of minutes corresponding to each part is easily calculated. The vernier usually contains lines of different lengths. There are as many of the longest lines as there are minutes in the smallest main-scale division. These lines are not usually so close together as the main-scale divisions, but by paying attention simply to the *number* of the long line which coincides most nearly with some main-scale division, we find the number of minutes to be added to the reading of the main scale (see § 40). Between the long lines, shorter lines are frequently placed, to represent fractions of a minute. Since a setting made by the eye, unaided by the telescope, is hardly accurate to a minute,[1] the student is advised to disregard these lines until he has mastered the reading of the sextant to degrees and minutes.

In angular as in linear measure, there is danger of making a mistake of a whole main-scale division (¶ 50, II.). If the reading of the main scale is thought to be about $x°$, and the vernier shows it to be a whole number plus 1', we record this reading as $x°\ 1'$; but if the vernier indicates a whole number plus $\overline{59'}$, we record the reading, not as $\overline{x°\ 59'}$, but $\overline{(x-1)°\ 59'}$.

---

[1] A man four miles off would subtend an angle of about one minute. A minute corresponds to a distance of less than one three-hundredth of an inch on a piece of paper held at the ordinary distance (10 inches) from the eye.

The first degree-mark below zero is counted as *minus one;* the second, *minus two*, etc. The number of minutes is always positive, since the vernier is made to read this way. To avoid confusion, the negative sign is written over the number of degrees, which it alone affects (see ¶ 50, I.). Thus, a negative angle of — 21' would be recorded $\overline{1}°$ 39'.

¶ 123. **Determination of the Zero-Reading of a Sextant.** — After a sextant has been adjusted as accurately as possible (see ¶ 122), its zero-reading must be determined. The index is first set at the zero of the main scale (as in ¶ 122, 5), the dark glasses are pushed out of the way, and the tube or telescope (c) is directed toward some distant object, — the smaller and brighter the better. A star is universally conceded to be the best object, but a distant electric arc-light will do. In the day-time, a church spire or the top of a flag-pole may answer. At sea the horizon line is frequently employed; in this case the plane of the sextant must be vertical. The angle between the mirrors should be so slight that the direct and doubly reflected images of the given object may at least be included in the same field of view. These images are then made to coincide by turning the tangent-screw (i, Fig. 106). Finally, the reading of the sextant is taken. This is called its zero-reading, because it corresponds to an angle, zero, between the direct and doubly-reflected rays.

It is easy to show that the fixed and revolving mirrors must be parallel when these rays (yb and xg, Fig. 107) are parallel; for the alternate interior angles b and e are equal by construction, hence their supplements, $a+c$ and $d+f$ must be equal. Now, the law of the reflection of light (§ 97) gives $a = c$, and $d = f$; hence, a being half of $a + c$ must be equal to d, which is half of $d + f$. Since c and d are alternate interior angles formed by the intersection of be with the mirrors ac and df, these mirrors must be parallel. Conversely, if the mirrors are parallel, the direct and doubly-reflected rays must be parallel.

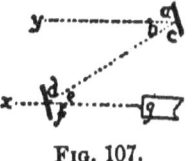

FIG. 107.

In order that the rays yb and xg may be sensibly parallel, let us say within one minute (1') of angle, the object from which they come must be 3,438 times as far off from the sextant as these rays are from each other. Since the perpendicular distance, bg, is generally less than a twelfth of a metre, it may be safe to employ any object more than 300 metres off for the determination of the zero-reading of a sextant with the unaided eye. To obtain results accurate to half a minute, the minimum distance must be doubled; for accuracy within 10" of angle the object should be at least 1,800 metres, or more than a mile away. For such results, a telescope (c, Fig. 106) must be employed.

¶ 124. **Determination of Small Angular Magnitudes by means of a Sextant.** — I. A sextant is to be set at or near its zero-reading; then turned so that the telescope (c, Fig. 106) may point directly toward the sun. The sextant is to be held so that its graduated arc may be in a vertical plane, below the revolving mirror (*b*). A sufficient number of dark glasses must be interposed in the paths of both the direct and the reflected rays. It is well to select these glasses so that the two images of the sun may differ in color, and thus be easily distinguished. The tangent screw (*i*, Fig. 106) is to be turned until the doubly-reflected image (R. I., Fig. 108) appears to be tangent to the direct image (D. I.) and below it, as in (1). The vernier is then read. Next, by turning the tangent screw the other way, the reflected image (R. I.) is made to move completely through the direct image, until it is tangent to, and above it, as in (2). The vernier is again read.

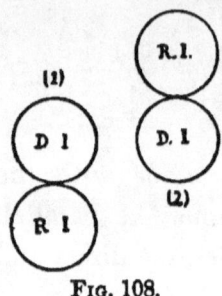

Fig. 108.

The first reading should be positive, the second negative. The average of the two should be found and compared with the zero-reading previously determined, with which it should agree. If the difference exceeds 1', the measurements in ¶¶ 123 and 124 should be repeated.

The second reading is now to be subtracted (algebraically) from the first. The difference, divided by 2, is evidently the angular diameter of the sun. The semi-diameter, which is quoted in all nautical almanacs, varies from month to month, according to the earth's distance from the sun. Its mean value is not far from 16'.

II. The sextant may also be used for the determination of the angular diameter of small terrestrial objects. The plane of the graduated arc must be held in all cases so as to be parallel to the diameter which it is desired to measure. The object should be so small that a negative [1] as well as a positive reading may be obtained, as in the case of the sun. The average of the two readings should agree with a zero-reading obtained from the *same object*, or from one at an equal distance. The difference between the two readings is not affected by parallax, since the error in both readings is the same. This difference, divided by 2, is therefore the angular diameter of the object in question as seen from the pivot of the revolving mirror. The position of this pivot should be noted, or the results will have no meaning. It is well for the student to measure either the actual diameter or the distance of the object in question, still better, both of these quantities; for though either may be calculated from the other, the two together give him the means of testing his inferences as to the manner in which his sextant should be read and an opportunity of confirming his results.

EXPERIMENT XLV.

PRISM ANGLES.

¶ 125. **Determination of the Angles of a Prism.**—I. A small prism (*abc*, Fig. 109) is fastened to the revolv-

---

[1] A sextant should be capable of giving negative readings down to 3, 4, or even 5 degrees.

ing mirror (*bc*) of a sextant with its axis parallel, as nearly as possible, to that about which the mirror turns.[1] The mirror is then rotated so that the direct image of a distant object, seen in the direction *ed*, may coincide with the image of the same object reflected first by the face of the prism (*ac*) then by the fixed mirror (*df*). If the two images cannot be made to coincide, the face *ac* is probably not parallel to the axis of the mirror, and must be made so by tilting the prism either from *b* to *c*, or from *c* to *b*, without separating the two faces, *bc*, of the prism and of the mirror. When parallelism is established, an exact coincidence of the images may be brought about. A reading of the sextant is then made. This serves to determine the prism angle *c*. In the same way the other two angles are determined.[2]

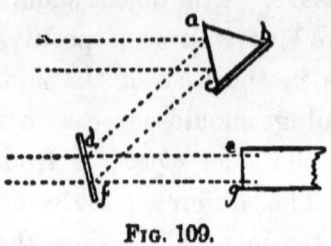

Fig. 109.

Subtracting from each reading of the sextant its zero-reading, determined as in ¶ 123, we have the indicated value of the angle corresponding to *c* (or *acb*) in the figure; for it is evident that the mirror *ob* in rotating from its zero position, *ca*, to the position

---

[1] The plane of the face, *ac*, should strictly pass through the axis of the mirror, to avoid errors of parallax. In practice, however, it is more convenient to mount the prism as in Fig. 109.

[2] To measure the three angles of a prism, one of which must be at least 60°, a sextant reading to 120° will be required. "Octants" are sometimes graduated to 120°; but do not read generally to more than 100°, on account of the space occupied by the vernier.

cb, turns through the angle *acb*. What we want, however, is the actual value of this angle, not the deviation of a ray of light striking the revolving mirror, which plays no part in the measurement. If, therefore, the sextant is found, as in ¶ 122, 7, to be graduated in half-degrees, half-minutes, etc., the indicated value of the angle must be halved in order to find the real value of *acb*.

The sum of the three prism angles should be 180°. A discrepancy of one or two minutes may be attributed (1) to errors of observation, (2) to pyramidal convergence of the sides of the prism, and (3) to errors in the adjustment or graduation of the sextant. If the measurements are several minutes in error, they should be repeated. If the same result is obtained, the parallelism of the prism faces should next be tested with a three-pointed caliper. With a perfect equilateral prism, we have evidently the means of detecting any error in the location of the 60° mark (or that numbered 120°).

II. Instead of a sextant, a spectrometer may be used, as will be explained in ¶ 126.

EXPERIMENT XLVI.

ANGLES OF REFRACTION.

¶ 126. **The Spectrometer.** — A spectrometer consists essentially of two telescopes (*ab* and *fg*, Fig. 110) capable of revolving about the centre of a graduated circle (*cde*). The eye piece of the first telescope is generally removed, and a narrow slit (*a*, Fig. 110) is usually substituted for the cross-hairs (*c*, Fig. 101, ¶ 116).

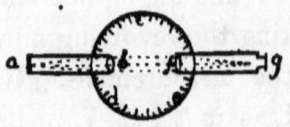

Fig. 110.

This slit is always at right-angles with the graduated circle, and at a distance from the lens, *b*, equal to its principal focal length; so that the rays from it may be rendered parallel by this lens (see ¶ 116, 3). The combination (*ab*) is called the "collimator" of the spectrometer. The telescope *fg* is focussed for parallel rays (¶ 116, 3), and carries an index with a vernier, by which its position on the graduated circle may be accurately determined.

A zero-reading can be found by pointing the telescope toward the collimator as in Fig. 110, and adjusting it so that the image of the slit, *a*, may be visible in the centre of the field of view, which is determined by the intersection of cross-hairs.

Let us now suppose that it is desired to measure the angles of a prism. The latter is mounted as in Fig. 111 (*cde*), so that the face *ce* may reflect part of the light from *ab* in the direction *fg*, and so that at the same time the face *cd* may reflect light in the direction *f'g'*. The telescope is then set so as to receive first one, then the other of the images of the slit, thus formed, in the middle of its field of view, and in each case a reading of the vernier is made.

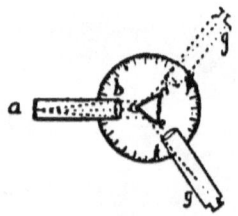

Fig. 111.

Let us suppose that the collimator is permanently set at 0° (or 360°) of the circle; that fg is at $x°$ and $f'g'$ at $y°$ of the circle. A radius of the circle perpendicular to ce would halve the angle x, on account of the law of reflection (§ 97); and hence would meet the circle at a point $\frac{1}{2} x°$. In the same way a radius perpendicular to cd would meet the circle half-way between $y°$ and 360°; or at $\frac{1}{2} y° +$ 180°; hence if prolonged backward it would meet the circle at $\frac{1}{2} y°$. Now, the angle between two surfaces may be measured by the angle between two lines perpendicular to them; hence the difference between $\frac{1}{2} x°$ and $\frac{1}{2} y°$ measures the prism angle dce. In other words, the angle between two faces of a prism is equal to half the angle between the two directions in which they reflect parallel rays of light. (Compare ¶ 120, 7.)

The most important adjustments of a spectrometer are the accurate levelling and focussing of the telescope and collimator for parallel rays (see ¶ 116, 3). The faces of the prism must be made perpendicular to the plane of the graduated circle as in ¶ 125. An instrument especially adapted to measure the angle between two reflecting surfaces is sometimes called a goniometer.

¶ 127. **Determination of Angles of Refraction.** — I. The telescope (fg), and collimator (ab) of a spectrometer are slightly inclined as in Fig. 112, so that a spectrum (¶ 128) of the slit, a, may be formed in the telescope by a prism dce,

FIG. 112.

the angles of which have been determined (¶¶ 125, and 126). The angle *c*, causing the refraction, should be placed symmetrically with respect to the telescope and collimator. If *dce* is an equilateral prism, an image of the slit may also be formed in the telescope by reflection from the face *de*. It is found that when the faces *cd* and *ce* are as stated equally inclined to the rays *ab* and *fg*, the angle between these rays reaches a minimum.

To make sure that this position has been approximately found, the prism should be rotated a little. The violet of the spectrum should be replaced by blue, green, yellow, and red, until finally the spectrum disappears altogether. It should make no difference whether the prism is turned to the right or to the left. If the spectrum moves in opposite directions when the prism is turned in opposite directions, the desired position has not been found. In this case the rotation should be continued in one direction or the other until the spectrum seems to come to a standstill. The prism is then very nearly in its "position of minimum deviation."

The slit should now be illuminated with light from a sodium flame,[1] the reflected image if necessary cut off, and the telescope roughly set on the yellow refracted image of the slit. Then the prism is turned slightly so that this image may move as far as possible towards the red (or less refrangible) end of the spectrum. The telescope is again set on the yellow image

---

[1] A common Bunsen burner beneath a netting of fine iron wire sprinkled with nitrate of soda furnishes an excellent "sodium flame."

more carefully than before, and the prism turned first to the right, then to the left, so as to find if possible a position in which the yellow image is even less refracted than before. Thus by successive approximations, the telescope may finally be set upon an image of the slit formed by the prism in its position of minimum deviation.

Subtracting the zero-reading (¶ 126) of the telescope from its reading when set upon the refracted image, we have finally the angle of minimum deviation in question; that is, the least angle through which sodium light may be bent in passing through the prism angle, dce, in the figure.

The relation between angles of refraction and indices of refraction is considered in § 102.

In repeating the experiment, the prism should be rotated through 180°, so that the rays would be bent upward instead of downward as in the figure. If the position of the collimator is unchanged, any error in the zero-reading may be eliminated (see § 44) by averaging the result with that previously obtained.

II. Instead of the spectrometer a sextant may be employed for the determination of angles of refraction. The prism is to be mounted as in Fig. 113, so that a ray of light from a distant point may be refracted by the prism angle c, previously de-

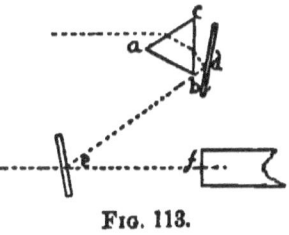

Fig. 113.

termined (¶ 125), then reflected by the revolving mirror d, and by the fixed mirror e into the telescope, f, where it is made to coincide with the direct ray, *ef*, from the same object. To obtain accurate results, monochromatic light should be employed; but a mean index of refraction may be found by making the direct image of a flame coincide with the yellow or green of its spectrum (§ 128). The prism must be placed by trial in the position of minimum deviation as with the spectrometer. The angle of deviation, being twice the angle between the mirrors, is indicated directly by the reading of the sextant, after the zero-reading has been subtracted.

The use of the sextant for the determination of angles of refraction is recommended only to those who have some skill in physical manipulation. For this reason a detailed description of the experiment has not been given.

¶ 128. **Spectra formed by the Dispersion of Light.** — The rays of light from a sodium flame, when bent (as in ¶ 127) by a prism, produce, with ordinary apparatus, a single yellow image of the flame. A flame colored with lithium gives similarly a red image, and one colored with thallium a green image. These images are not, however, in the same direction from the observer, owing to the fact that rays of different hues are unequally bent by a prism. Indeed, if a flame be colored by a mixture containing certain proportions of lithium, sodium, and thallium, three images of the flame — one red, one yellow, and one green — may be seen side by side, distinctly separated by dark spaces between them. Many substances, even when chemically pure, cause under the same circumstances several distinct images of a flame to be produced. Each of these images differs in hue from the rest. The images may be more or less bright and more or less widely separated. Together they constitute what is called the *spectrum* of the substance producing them. When, as in a common gas-flame, light of every hue is represented, an indefinite number of images are formed, and these necessarily overlap one another. The result is called a continuous spectrum.

An instrument intended simply to examine spectra with a view to observing the number of images present, is called a spectroscope. An instrument like that described in ¶ 126, especially adapted to the determination of angles of refraction, through settings made upon the differently colored images in a spectrum, is properly called a spectrometer.

Those substances which bend light the most usually produce the greatest separation or "dispersion" of rays of different colors. There is, however, no definite proportion between the effects of refraction and dispersion. Thus an equilateral prism of crown glass which bends rays of light about 40°, separates the extreme red and violet rays by about 4°; while a prism of flint glass, producing nearly double the dispersion, bends rays less than 50°.

To determine the dispersive power of a given substance, two indices of refraction are generally found (see § 102), one with red light, the other with violet light. The red light selected is of a peculiar wavelength (§ 98), namely, .00007604 cm., being that which causes the line A in the solar spectrum. The violet light has similarly a wave-length .00003933 $cm.$, corresponding to the line H_2 of the solar spectrum. The difference between the indices of refraction of a given substance for these two rays is sometimes called the "index of dispersion" of the substance in question.

EXPERIMENT XLVII.

WAVE-LENGTHS.

¶ 129. **Theory of the Diffraction Grating.** — When a distant candle is looked at through a linen handkerchief, or through any fine network, several images of the candle are usually seen (Fig. 114). These are not, however, as one is at first apt to suppose, simply so many views of the candle through the meshes of the handkerchief; for each image represents the whole candle, and the distance between the images is not only disproportionate to the size of the meshes, but actually increases as the meshes become smaller. It is, moreover, unaffected by the distance of the handkerchief from the eye. The phenomenon is an example of *diffraction* (§§ 100, 101), and depends upon a re-

Fig. 114.

lation between the length of the waves of light and the distance between the threads.

The central image (*a*, Fig. 114) is the direct image of the candle. It may be distinguished from the side images, *a'* and *a"*, for instance, both by its greater distinctness and by the absence of color.

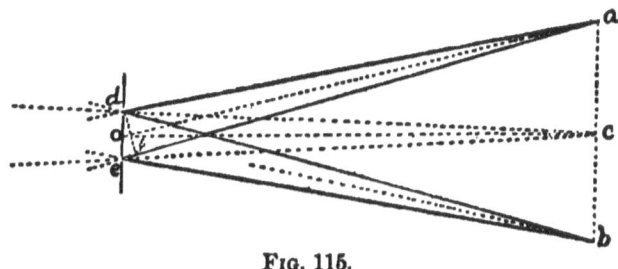

Fig. 115.

The side images will be found tinged with blue on the side toward *a*, and with red on the outer side. This is due to the fact that different colors are unequally bent by diffraction. Each of the side images is in fact a "spectrum" of the candle. It is interesting to place two candles at points corresponding

Fig. 116.

to *a* and *b* (Fig. 115) at such a distance that when the candles are viewed through a network *de*, the side image at the right of *a* may coalesce with the side image at the left of *b*, so as to form a single image at the point *c* similar to that represented in

Fig. 116. If *o* is one of the threads, *d* and *e* the spaces between it and the two parallel threads on either side of it, then drawing *ad*, *ao*, *ae*, *cd*, *co*, *ce*, etc., also *df* perpendicular to *ao*, we have (since *ao* practically bisects the angle *a*) $ad = af$. The path *ae* is accordingly longer than *ad* by the distance *ef*, which must therefore be the length of a wave of light (§ 101), since the rays do not interfere. Now, by similar triangles, we have,

$$ef : de :: ac : ao; \qquad \text{I.}$$

hence, if we know the distance, *de*, between the threads, the distance, *ao*, between the handkerchief and one of the candles, and the distance, *ab*, between the candles, so that by halving the latter the distance of the side image (*ac*) may be found, we may calculate the average length (*l*) of a wave of light by the formula,

$$l = \frac{de \times ac}{ao}. \qquad \text{II.}$$

The student may himself estimate wave-lengths in this way. For a human eye in its normal condition (see ¶ 115) the average wave-length in a candle-flame has been found to be about 60 millionths of a centimetre. We may make use of this fact to estimate the distance (*d*) between the threads of the handkerchief by the formula derived from I.,

$$d = .00006 \times \frac{ao}{ac}. \qquad \text{III.}$$

The angle *aoc* is called the angle of diffraction The ratio *ac* : *ao* is by definition the sine of this angle;

hence if the angle be measured, the ratio can be found from Table 3.

¶ 130. Determination of Angles of Diffraction.— An ordinary diffraction grating (see § 101) consists of a set of parallel and equidistant lines ruled or photographed on glass (Fig. 117). A candle-flame viewed through such a grating gives several images, as in the case of a netting (Fig. 114); but these images are all in a single row (Fig. 116). The relation between the wave-length, distance between lines, and angle of diffraction is the same as in the case of a netting (¶ 129). The angle of diffraction may be determined either by a sextant (¶ 124), or by a spectrometer (¶ 126). In any case the lines of the grating must be perpendicular to the graduated arc or circle by which this angle is to be determined.

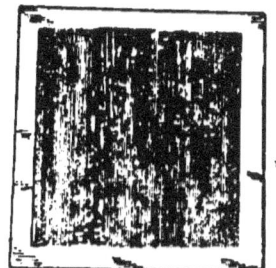

Fig. 117.

I. A coarse diffraction grating, containing from 10 to 20 lines to the millimetre, is to be mounted directly in front of the tube or telescope of a sextant (c, Fig. 106, ¶ 121), which is then to be pointed at a distant sodium flame (¶ 127). When the fixed and revolving mirrors are nearly parallel, it should be possible to see the flame (either directly or by double reflection) with at least two images due to diffraction, one on each side of it (see $a'a''$, Fig. 114). The experiment consists in measuring the angular distance between the two side images next the flame, by the method already explained in ¶ 124.

Let *a* and *b* (Fig. 116) represent the direct and doubly reflected images of the flame. The revolving mirror is first set so that the side image at the left of *b* coalesces with the side image at the right of *a*, to form a compound image, *c*, as in the figure. Then the image (b'') at the right of *b* is made in the same way to coalesce with the side image (a') at the left of *a*. The two readings are then subtracted algebraically, one from the other, and the result is divided by 2 (as in ¶ 124), to find the angle subtended by the side images (a' and a'', Fig. 114). This angle must again be divided by 2 to find the angular distance of either of the side images from the flame. This angular distance evidently corresponds to the angle *aoc* (Fig. 115), and is, accordingly, the angle (*a*) of diffraction in question.

Since the wave-length of sodium light is .0000589, we have, substituting this value in formula III., ¶ 129, for the distance (*d*) between two lines of the grating.

$$d = \frac{.0000589}{\sin a} \qquad \text{I.}$$

Fig. 118.

A grating, thus tested, serves as a convenient scale by which the diameters of small objects may be determined. Such a scale is interesting, because it represents the nearest approach to an absolute standard of length (see § 5).

II. Instead of a sextant, a spectrometer may be employed to measure angles of diffraction. If the grating is mounted in the centre of the graduated circle (Fig. 118), so as to be perpendicular to the collimator, ab, the reading of the telescope, fg, when set upon one of the side images, will determine the angle of diffraction in question. It is not very easy, however, to make the grating accurately perpendicular to the collimator, and the slightest deviation affects the angle of diffraction. A grating, like a prism (see ¶ 127) is found to have a position of minimum deviation, when it is equally inclined to the direct and diffracted rays (see de, Fig. 118). This position may be found by trial in the same way as with a prism.

When the method of minimum deviation is employed, the formulæ of ¶ 129 must be somewhat modified.[1]

The wave-lengths contained in Table 41 were determined by a method essentially the same as the one here given.

[1] In Fig. 115 each ray is supposed to lose one wave-length with respect to the next before reaching the grating. If, however, the grating is equally inclined to the incident and diffracted ray, the loss must be half a wave-length before, and half a wave-length after reaching the grating; that is, $ef = \frac{1}{2} l$. The angle aoc will represent also half the total angle of diffraction; or $aoc = \frac{1}{2} a$. If d is the distance de between the lines, we have, substituting $\sin \frac{1}{2} a = \sin aoc$ for $ac \div ao$ (see ¶ 129), and multiplying by 2,

$$l = 2d \sin \tfrac{1}{2} a.$$

EXPERIMENT LXVIII.

INTERFERENCE OF SOUND.

¶ 131. **Determination of the Wave-Length of a Tuning-Fork by the Method of Interference.** — I. The two ends of a thick-sided rubber tube, about half a metre long, and with an internal diameter of at least 5 *mm.*, are joined together, as in Fig. 119, by a Y-joint, and a tube connected with the stem of the Y is held to the ear. A tuning-fork making from 400 to 600 vibrations per second (as for instance a "violin A-fork" or a "C-fork" just above it) is then touched lightly to the tube at different points, as in the figure. The note emitted will generally be plainly heard; but two or more points will be found at which the sound is nearly extinguished. These points are to be marked with ink on the rubber tube. Then the tube is to be disconnected from the Y-joint, straightened out, but not stretched, and the distance between adjacent marks carefully determined by a metre rod.

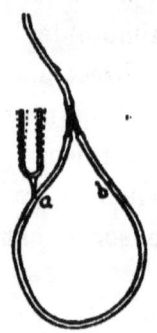

Fig. 119.

The extinction of the sound is due to the interference of vibrations reaching the Y-joint by the two different channels (§ 100), which differ either by half a wave-length, or by some odd multiple of half a wave-length. It follows that two adjacent points, *a* and *b* (Fig. 119), where the sound reaches a minimum, must be half a wave-length apart. To find the length of a wave of sound created in the tube by the vibration of the tuning-fork in question, we have therefore only to multiply the distance *ab* by 2.

Wave-lengths depend more or less upon the temperature of the air in the tube, which should therefore be noted. They are generally less in small tubes than in the open air, particularly if the sides of the tube be yielding. The interference is never complete, because the wave which travels the longer distance becomes weaker than the other, and hence cannot wholly destroy it. The points where the sound reaches a minimum may often be located more exactly when a fork is vibrating feebly than when it is sounding loudly.

II. In place of a rubber tube, we may employ a pair of telescoping U-tubes (Fig. 120), forming a closed circuit. Near the junctions two openings are made. One of these is connected with the ear, the other receives vibrations propagated from a tuning-fork through the air. The two channels by which the sound reaches the ear may be made unequal in length by drawing out the tubes. The difference between them may be measured by graduations on the inner tube, or in any other obvious manner.

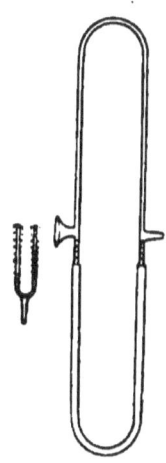

Fig. 120.

The smallest difference between the two channels which can produce interference is half a wave-length; hence, we multiply it by 2 to find the wave-length in question.

From the wave-length of a fork in air, we may calculate roughly its rate of vibration (¶ 134, formula II.).

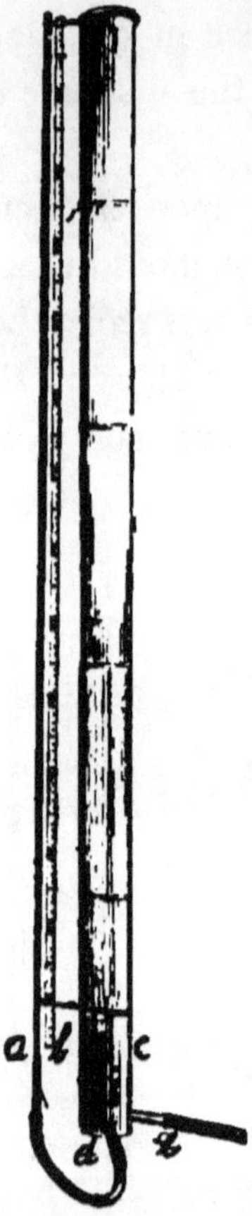

EXPERIMENT XLIX.

RESONANCE.

¶ 132. Determination of Wave-Lengths by the Method of Resonance. — A metallic tube or "resonator" $1\frac{1}{2}$ metres long and 10 $cm.$ in diameter (c, Fig. 121) is filled with water; then a tuning-fork, making from 200 to 300 vibrations per second, is held near the mouth of the tube, while the water escapes by the spout, e. When the water falls to a certain level, the note emitted by the fork, instead of dying away, will suddenly swell out. The flow of water is then checked. Water from the faucet is now admitted to the resonator by the spout e, and again allowed to escape, with a view to finding at what level it gives the maximum resonance. The variation in the loudness should be observed both when the water is rising and when it is falling. By alternately increasing and diminish-

Fig. 121.

ing the quantity of water in the tube, the desired level may be located within a millimetre. This level is then read by the gauge ab, consisting of a millimetre scale, b, and a glass tube, a, connected by a rubber tube (d) with the resonator.

The fork is now kept in vibration while the level of the water is allowed to fall to a much greater depth than before. A second point of resonance is thus located in the same way as the first. The temperature of the air within the tube should be carefully noted.

The distance between the two points of maximum resonance is found by subtracting one scale-reading from the other. This distance is (see § 99) exactly half a wave-length, and hence must be multiplied by 2 to find the wave-length of the fork.

The rate of vibration of the fork may now be calculated approximately, as will be explained in ¶ 134, by formula II. of that section.

EXPERIMENT L.

MUSICAL INTERVALS.

¶ 133. **Determination of Musical Intervals.** — I. METHOD OF INTERFERENCE. — The wave-lengths of two forks are to be determined as in Experiment 48, taking care that the temperature of the air is the same in both cases, and the musical interval between the forks is to be calculated as in ¶ 134, III.

II. METHOD OF RESONANCE. — Instead of using the method of interference, we may determine the wave-lengths of two forks by the method of resonance, as in Experiment 49, with care as before to avoid changes of temperature. The musical interval should be calculated in the same way (¶ 134, III.).

III. PYTHAGOREAN METHOD. — An instrument which will be found convenient for the determination of musical intervals is represented in Fig. 122. It is

FIG. 122.

called the "*monochord*," and is attributed to Pythagoras. In modern instruments, it consists of a steel wire, *fbcdeg*, fastened to a board at *g*, then passing over two "bridges" (or triangular supports, *e* and *b*) round a pulley (*f*) to a weight (*h*) by which it is kept stretched with a constant force. The positions of the bridges are determined by a graduated scale.

The wire (*be*) is set in vibration by a bow (*ai*), and the distance between the bridges (*b* and *e*) is varied until the note emitted by the wire is in unison with one of the forks. The distance (*be*) is then adjusted so as to produce unison with the other fork. From the two distances in question, the interval between the forks is to be calculated as in ¶ 134 (Formula III.).

Determinations with a monochord should be attempted only by students having a more or less musical ear. The exact adjustment of two notes in unison may be inferred from the cessation of "beats" (Exp. 53).

IV. HARMONIC METHOD. When the musical interval between two forks has been determined by any of the preceding methods, or simply recognized by the ear, the exactness of the interval in question may be tested as follows: The bridges b and e (Fig. 122) are first placed at a distance which is the *least common multiple* of the two distances giving unison with the two forks. By touching the string lightly with a feather (c, Fig. 122) at certain points, it may be made to vibrate in segments as in the figure. The number of segments is first made such that the string is nearly in unison with one of the two forks, and the distance (de) adjusted if necessary so that the unison may be perfect. If the wire can be made to divide in such a manner as to sound in unison with the other fork, there must be an exact musical interval between the forks. If, on the other hand, beats are heard, the interval is probably inexact, and by an amount which may be estimated from the frequency of the beats (Exp. 53).

For the practical application of this method, the monochord should be capable of giving a very low note, at least two octaves (¶ 134) below the lower fork; hence the tension of the wire must not be too great. The lowest note which a string can give out under given circumstances is called its "fundamental tone." The other tones are caused by its division into segments, separated by still points or "nodes." These tones are called the "harmonics" of the string. The musical interval between any two harmonics may be calculated from the number of vibrating segments (see ¶ 134, IV.), which must therefore be noted in each case.

¶ 134. Theory of Musical Intervals. — If a tuning-fork gives out n waves each l centimetres long in one second, then the furthest wave must be nl centimetres off from the fork at the end of that space of time; and since it travels nl cm. in 1 sec., the velocity of sound must be nl cm. per sec. The fundamental equation connecting the number (n) of vibrations per second, the wave-length (l), and the velocity of sound (v) is, therefore, —

$$v = nl. \qquad\qquad \text{I.}$$

The velocity of sound in air of any temperature may be found from Table 15 B. If the humidity is unknown, a mean value (60 per cent) may be assumed; then if the wave-length of a given fork is l, we have, —

$$n = \frac{v}{l}. \qquad\qquad \text{II.}$$

When two forks give n' and n'' vibrations per second, with wave-lengths respectively of l' and l'' centimetres, we have from II., —

$$n' = v \div l', \qquad (1)$$
and
$$n'' = v \div l''; \qquad (2)$$

hence, dividing (1) by (2),

$$n' : n'' :: l'' : l'. \qquad\qquad \text{III.}$$

The ratio of the rates of vibration is called the musical interval between the forks, and is accordingly in the inverse ratio of their wave-lengths.

Formula III. is applicable to a wire as well as to a tube. When a wire of the length be divides into N segments, the length of each must be $be \div N$; we have accordingly for the lengths l' and l'' of the segments formed by the division of the wire (be) into N' and N'' parts, respectively,—

$$l' = be \div N', \quad (3)$$
$$l'' = be \div N''; \quad (4)$$

hence, dividing (4) by (3),

$$l'' : l' :: N' : N'', \quad (5)$$

which, substituted in III., gives

$$n' : n'' :: N' : N''. \quad \text{IV.}$$

This shows that the rates of vibration of different harmonics are proportional to the number of vibrating segments in the wire.

It has been stated that the ratio between two rates of vibration, n' and n'', determines the interval between the two notes to which they correspond. The ordinary musical scale consists of a series of notes whose rates of vibration, whether high or low, are always relatively proportional to the following numbers set beneath their names:—

DO	RE	MI	FA	SOL	LA	SI	DO
24	27	30	32	36	40	45	48

The interval between the first and third note of this series is called a "third;" between the first and fourth, a "fourth," etc. The first two are said to be one tone apart; the last two, one semitone apart. The most common musical intervals may be arranged as follows, according to the simplicity of the ratios which they involve when reduced to their lowest terms: —

Name.	Ratio.	Name.	Ratio.	Name.	Ratio.
Unison	1:1	Fourth	4:3	Minor Third	6:5
Octave	2:1	Sixth	5:3	Whole Tone	9:8
Fifth	3:2	Third	5:4	Semitone	16:15

The sum of two or more intervals is always represented by the product of the ratios in question; thus, when we say that two notes are an octave and a fifth apart, we mean that the higher makes one and one half times as many vibrations per second as the octave of the lower note; or, again, twice as many vibrations as a note a "fifth" above the lower note; that is, in either case, three times as many vibrations as the lower note itself. In the same way an interval of two octaves corresponds to the ratio 4:1 between the rates of vibration; an interval of three octaves corresponds to the ratio 8:1, etc. It is a fact to be noted that the musical intervals involving the simplest ratios are the most agreeable to the ear.

END OF PART FIRST.

www.ingramcontent.com/pod-product-compliance
Lightning Source LLC
Chambersburg PA
CBHW031331230426
43670CB00006B/316